KNOCK ON WOOD

KNOCK
ON
WOOD

Luck, Chance, and the Meaning of Everything

JEFFREY S. ROSENTHAL

HARPER **PERENNIAL**

Knock on Wood
Copyright © 2018 by Jeffrey S. Rosenthal.
All rights reserved.

Published by Harper Perennial, an imprint of HarperCollins Publishers Ltd

First published by HarperCollins Publishers Ltd in a hardcover edition: 2018
This Harper Perennial trade paperback edition: 2019

HarperCollins books may be purchased for educational, business,
or sales promotional use through our Special Markets Department.

HarperCollins Publishers Ltd
Bay Adelaide Centre, East Tower
22 Adelaide Street West, 41st Floor
Toronto, Ontario, Canada
M5H 4E3

www.harpercollins.ca

Library and Archives Canada Cataloguing in Publication
information is available upon request.

ISBN 978-1-4434-5308-0

Printed and bound in the United States
LSC/H 9 8 7 6 5 4 3 2 1

In memory of my mother, Helen S. Rosenthal

Contents

Chapter 1   Do You Believe in Luck? 1
Chapter 2   Lucky Tales 6
Chapter 3   The Power of Luck 14
Chapter 4   The Day I Was Born 24
Chapter 5   Our Love of Magic 33
Chapter 6   Sharpshooter Luck Traps 48
Chapter 7   Luck Revisited 55
Chapter 8   Lucky News 67
Chapter 9   Supremely Similar 82
Chapter 10   Interlude: The Case of the Haunted House 91
Chapter 11   Protected by Luck 105
Chapter 12   Statistical Luck 116
Chapter 13   Repeated Luck 131
Chapter 14   Lottery Luck 146
Chapter 15   Lucky Me 159
Chapter 16   Lucky Sports 173
Chapter 17   Lucky Polls 184
Chapter 18   Interlude: Lucky Sayings 192
Chapter 19   Justice Luck 203
Chapter 20   Astrological Luck 214
Chapter 21   Mind over Matter? 235
Chapter 22   Lord of the Luck 250
Chapter 23   Lucky Reflections 264

Acknowledgements 273
About the Author 275
Glossary 277
Notes and Sources 279
Index 325

KNOCK ON WOOD

CHAPTER 1

Do You Believe in Luck?

As a university professor who specializes in the mathematics of probability and statistics, I am dedicated to spreading knowledge and wisdom about the workings of randomness and uncertainty. I have confidently answered questions about all sorts of probability-related subjects—lotteries, airplane safety, election polls, crime rates, gambling odds, sports statistics, medical testing, and more. But then, every once in a while, someone asks me if I believe in luck. An awkward pause follows, and then I try to eke out a reply.

Do I believe in luck? Well, sure, of course I do. Sometimes things work out well, and sometimes they don't. Sometimes external forces make life difficult, and other times they come together just right. In my own case, I was lucky enough to be born into a middle-class family that valued education and started me on the path to success. I was lucky enough to grow up in a peaceful, safe, prosperous country. I was lucky enough to be admitted to top universities, leading to a good academic position with the job security of tenure. Of course I believe in luck!

Do I believe in luck? No, certainly not. Some people believe in such things as unlucky numbers, astrological predictions, and lucky charms, which all seem like bunk. There are no known physical laws that could produce a causal connection between actual life outcomes and any of these unusual items, nor have careful experiments ever shown any consistent relationship between them. Seriously believing in any of them seems a bit absurd. And just because good things have happened to someone so far does not mean that pattern will continue. The past does not predict the future, patterns are not set, and no one is guaranteed any luck at all. Of course I don't believe in luck!

Do I believe in luck? In the end, it depends what you mean. Luck is one of those words that can be interpreted in many different ways. I once did a radio interview where they asked me to begin by giving a simple, short definition of luck.[1] I soon discovered that I couldn't do that, and the entire interview got bogged down in a debate about what we were actually talking about.

If you say something happens "by luck," then it is clear you are *denying* that it happened due to established scientific cause and effect (like a ball falling to the ground because of gravity), or due to hard work (like passing your final exam because you studied hard), or due to specific intent (like a bucket of water dumping on your head because your goofy friend left it balanced on top of your door). But if that is what luck isn't, then what is it that luck *is*?

Sometimes people use the word "luck" to simply describe events that are outside of our control or prior knowledge, a sort of *dumb luck* or *random luck*. We cannot predict such luck, only notice it in hindsight—for example, you go to the store to buy sneakers and find that they are on sale this week, which you hadn't known or suspected before leaving your house. Or you're visiting a foreign city when terrorists attack, and you're relieved to hear that their bomb went off on the opposite side of town from your hotel. These are indeed instances of getting lucky, but only in the sense of benefiting from a situation you could not influence or predict in any way. And if that is all there is to it, then it's nothing but dumb, random luck.

On the other hand, sometimes people use the word "luck" to allude to certain special powers that magically affect the future—everything from lucky charms, like rabbits' feet and four-leaf clovers, to supernatural "predictions" made by horoscopes, fortune cookies, and tea leaves, to "fate" about what just *had* to happen, to "karma" taking its sweet revenge, to simply being a magically lucky guy to whom good things must always happen. All of this implies a sort of *forceful luck*, a special kind of luck that can be predicted in advance and affects the probability of future events, based not on scientific laws or hard work or other fact-based explanations, but rather on other, ethereal causes.

So, which one is correct? Does "luck" refer to something that is dumb and meaningless, or to something that is forceful and magical?[2]

So many people believe in a special luck force, in one form or another. They scoff at my usual "scientific" approach to randomness and luck, incredulous that I could possibly imagine that probability and scientific causes are all there is. Could they be right and I be wrong? How can we truly measure and evaluate luck? How can we decide which predictions are accurate and which are nonsense? How can we figure out what causes what? How can we determine what really does govern the randomness all around us?

And how could I sort this all out? Well, maybe by writing a book.

In the pages that follow, I will discuss various examples of luck at work and try to sort out luck's meaning—or lack thereof. Some of the questions I will ponder include:

- Why am I, like so many others, attracted to fictional works like *Macbeth* and *Shoeless Joe* whose plot lines centre around concepts of luck, fate, and destiny? Why do we love magic in our stories? Are we rejecting scientific attitudes each time we read them? And should we then expect life to imitate art by following its own rules of destiny and magical meaning?

- If a friend tells me that sun rays shining through some tree branches were a "sign" to bring her comfort, is she onto something? Were those sun rays specially designed to cheer her up, or were they just random? And if they were random, does that make her comfort less real?
- When sports fans face the disappointment of their team losing, why are they so quick to blame it on superstitious curses? Does their belief fulfill some need? Is there any logic to their feelings?
- When terrible tragedies like nuclear explosions, hurricanes, and tsunamis kill thousands of people, is there a "reason" for it? Are the victims fulfilling their "destinies"? Or is it just dumb, horrible luck, causing suffering for no reason at all?
- How should I react to learning that I was born on a Friday the Thirteenth? Does that fact doom me to a life of failure and bad luck? Have my modest successes demonstrated that I "broke" the curse? Or was there never any curse to begin with?
- When evaluating my academic career success, should I feel proud of my accomplishments? Or should I just dismiss them all as meaningless, undeserved luck?
- When my student went on a blind date and discovered that both he and his date drove identical cars, was that a sign they were meant to be together? Did it guarantee love and bliss? Should it influence the decisions they make?
- What can we learn from amazing stories like the man who felt mysteriously drawn to a particular Hawaiian beach, where he happened to meet a half-brother he had never known, an encounter that ended up changing his life and rescuing him from his troubles? Was this destiny at work? Was it the hand of God? Or was it simply a

random, lucky coincidence that could just as easily never have occurred?

- Why do so many people believe in astrological horoscopes, psychic predictions, fortune tellers, numerology, and other supernatural phenomena? Is there any basis for them? Are there scientific studies that evaluate the evidence for them?
- When we buy lottery tickets, gamble at a casino, or roll dice in a board game, is the randomness fixed, or is it subject to our influences? Is there anything we can do to improve our luck? Are some people inherently luckier than others?
- When we read a news story about the latest medical study, a new poll, or an astounding coincidence, should we trust it? Does it have meaning? How can we distinguish what is truly significant from what occurs just by luck alone?
- And most important of all, how can we answer these questions? What principles can help us distinguish between random, meaningless, dumb luck and true instances of meaning and significance and influence and destiny? What *luck traps* must we watch out for, to avoid drawing the wrong conclusions? How can we identify causes when they arise, without imagining them when they are lacking?

These questions don't have easy answers. They have occupied me, and sometimes unsettled me, for years. My perspective is often at odds with those who surround me—when I am able to articulate my perspective at all. I had some misgivings about tackling these issues in book form, and yet here I am. Let the adventure begin.

CHAPTER 2

Lucky Tales

Our lives are constantly subjected to unexpected twists and turns that are beyond our direct control. Surprises that help us, hurt us, or confuse us. We can try to plan and prepare all we want, but the world may have other ideas for us. As the great Robbie Burns once put it:

> *The best-laid schemes o' mice an' men*
> *Gang aft agley,*
> *An' lea'e us nought but grief an' pain*
> *For promis'd joy!*[1]

Can these twists and turns be dismissed as just random luck? Are they the result of simple science at work? Are they rooted in probability, odds, and outcomes? Many people do not think so. They insist that these events are somehow controlled by powerful supernatural forces such as superstition, ESP, divine intervention, and destiny.

Luckily for me, I haven't had to search very hard to find examples to illustrate the latter point of view. They come up in my everyday conversations, with strangers, acquaintances, and friends alike.

The Scottish Play

A friend once hosted a party where each guest was supposed to recite a piece of poetry. I nervously selected a soliloquy from my favourite Shakespeare play, *Macbeth*. I chose the famous one that begins, "Is this a dagger I see before me?" which I had to memorize back in high school. I read it boldly, in my most ferocious Scottish accent, to moderate acclaim from my fellow partygoers. Not enough to quit my day job, but so far, so good.

Later, my friend made the mistake of bragging to another woman about the brilliance of my earlier *Macbeth* reading. The woman immediately became concerned. "You read *Macbeth* out loud?" she gasped. "But didn't you experience bad luck afterwards?" She was referring to the superstition that misfortune is somehow caused by quoting *Macbeth* (or even naming it; hence alternative names like "The Scottish Play" and "The Bard's Play" and "Mackers"). This superstition has its origins in a series of accidents that supposedly occurred during various productions of *Macbeth*, including a stabbing during a 1794 production.[2] Many of these misfortunes are poorly documented and difficult to verify, but that does not dissuade the true believers.

Surprised, I replied that no, I had not experienced any resulting bad luck. I mentioned, as mildly as I could, that I don't actually believe in such superstitions. But this only provoked her to say more. "Oh, my daughter didn't believe either," she persisted. "So she quoted *Macbeth* too. Then, a week later, while preparing for a trip, she discovered that her passport had expired and had to be renewed!"

I began to realize that the woman was actually serious. Indeed, she was *certain* that her daughter's passport woes were the result not just of simply random bad luck, but of some other, mysterious force. And nothing anyone said would change her mind. Why was she so convinced? Should I become convinced too?

Further enquiries revealed that the daughter's passport had expired two years earlier, after being issued years before that. So I politely asked the woman how its expiration, the result of actions from long ago, could have been caused by events from the previous week. "It is not a question of *cause*," she replied huffily as she walked off.

So much for my probabilistic diplomacy. Clearly this woman and I were coming from two very different perspectives. But who was right, and who was wrong? Is it even possible to say?

Disappearing Diamond

A friend once told me about the time, in the middle of a long drive, when she noticed that the diamond was missing from her wedding ring. Understandably upset, she searched the car high and low, without success. A religious person, she prayed for God to return the diamond to her. Some time later, she exited her car at a rest stop and the diamond emerged, unharmed, from a fold in her shirt.

This was proof, my friend assured me, of divine intervention. The fact that she had prayed for the diamond's return and subsequently found it showed that her prayers had been answered, that God had intervened to bring her diamond back to her. She couldn't understand how anyone could fail to believe in divine intervention after hearing her tale.

Considering the issue, I asked her why God had allowed her diamond to get lost in the first place. If God had really helped her recover the stone, would it not have been simpler to prevent it from vanishing at all? But she had a ready answer for that too. The diamond getting lost, she explained immediately, had been the work not of God, but of the devil.

So her conclusion from the story was clear. Was mine?

Craps Karma

I was surprised one day when I was invited onto a popular daytime talk radio program that wasn't known for its interest in science.[3] I was happy to comply, but it did make me wonder why they had invited me.

As the interview began, the reason for my invitation became clear. The show's host was fascinated by craps, the gambling game based on repeatedly rolling a pair of dice, whose complicated rules turn out to give the player a 49.29 percent chance of winning. And why was she so interested in that game? Did she, like me, find the probability aspect fascinating? Nope. Instead, she explained that craps involves lots of karma. I immediately realized that this would not be a typical interview about probability.

I nervously asked her to elaborate. She explained that she had attended a week-long craps "school" in Georgia, where she had learned all kinds of interesting "craps karma" rules.[4] For example, she had learned that when shooting craps, if the dice happen to accidentally fall on the floor, it is a sign that the shooter will probably lose on their next roll. So at that point you should bet heavily *against* the shooter (yes, this is allowed—it's called a "don't pass" bet) and reap the likely rewards.

I hemmed and hawed as best as I could. Back then, I hadn't done too many media interviews, and I didn't want to be seen to be contradicting my host. I said something about finding her perspective "interesting," though I mentioned that when it came to the karma theories she was explaining, I "might not completely agree."

But what was I *really* thinking?

Bracelet ESP

I once gave a public talk with the rather bold title "Why Statisticians Don't Believe in ESP." My presentation seemed to go very well. It even included an interactive card-guessing game, which I used to demonstrate that no one in the audience had any special extrasensory perception (ESP) that would allow them to consistently guess which card was which. I felt pleased that I had managed to make my points clearly and convincingly to the 100 or so listeners.

But after my talk was over, an audience member approached me and firmly declared, "You can't say that there is no ESP at all." I meekly replied that, well, actually, I was indeed saying that. She immediately continued, "Ah, but the other day, I lost a bracelet and couldn't find it anywhere. Then, that night, I had a vivid dream that it was in the dumpster behind my apartment. The next morning, I searched the dumpster, and sure enough, I found my bracelet, unharmed." She had no doubt that this was absolute proof of her psychic powers.

In a way, her story was quite compelling. Who among us has not, at some point, had a profound-seeming dream, and then awoken to find out that it offered great and unexpected insights or information? Dreams seem mysterious and magical at the best of times, so it is very easy to imagine that they create forceful luck.

I responded to the audience member by meekly mentioning that perhaps she had also had *other* dreams that had not worked out so well, which might perhaps undermine her claim of ESP. But a friend of mine—to my surprise—immediately jumped to her aid. "Why does ESP have to work *every* time in order to exist?" he demanded.

The combination of the audience member's certainty and my friend's support for her position was more than I could handle—especially right after giving a public talk. *Luckily*, the discussion soon turned to other matters, and we moved on. But in my head, the questions remained: Did the audience member really believe in ESP? Did my friend really agree with her? If someone occasionally learns something from a dream, is that convincing evidence that they have ESP? How should I have responded to all of this?

My Friend's Girlfriend's Lottery Ticket

Just after agreeing to write this book, I attended a local theatre festival, where I ran into an old friend and was introduced to his girlfriend. Hoping to make a good impression, I explained that I was a professor of statistics. To my pleasant surprise, she was delighted—a rather different reaction than the "Oh, I hated my statistics class" response I often receive. So far, this conversation was going great.

My friend's girlfriend then said she wanted to ask me something. I eagerly accepted.

"I sometimes buy lottery tickets," she began. Now, this wasn't the most promising of beginnings—I generally think lotteries are silly because the chance of winning is so low. But still, I was optimistic that she might be leading to an interesting question.

"You see," she continued, "I used to think in terms of the actual *numbers* I was picking." Um, okay, I thought, it's hard to argue with that. So in what terms did she think now?

"The last time I bought a ticket," she explained, "I thought more in terms of the *spaces* between the numbers I pick." Apparently she was referring to the other numbers on the lottery card, the ones she was *not* circling. So if she picked the numbers 14 and 18, then I guess 15 and 16 and 17 would be the "spaces."

"And I think it was really helpful: I matched four numbers out of six!"

I wasn't sure how to reply, so I very cleverly said nothing.

"Of course," she went on, "it's just intuition, and intuition can't be *proven*. But do you think it's a good idea? Or do you think my four matches were just a coincidence?"

Well, did I?

Blind Date

In a class discussion about randomness some years ago, one of my students told an interesting story. He was once set up on a blind date. On the appointed day, he drove to the designated restaurant, parked his car, and waited for his mysterious match to arrive.

After a number of other cars had arrived, one caught his eye. He could see that it was the exact same make and model as his own car—and the exact same colour too. On closer inspection, he determined that it was even manufactured in the same year. Intrigued, he wondered if this car could possibly belong to the lady he was about to meet.

Finally a woman emerged from the car, and he discovered that yes, this woman was indeed his date! Without any planning, he and his blind date had arrived at their rendezvous in identical cars. Surely this was fate! Destiny! Karma! Kismet! Surely this indicated that their date was "meant to be," and they would live happily ever after. Right?

Right?

Controlling Luck?

These stories all have one thing in common: an effort to control or explain the luck in our lives. Whether it is an expired passport or a lost diamond, a gambling defeat or a dumpster dream, a lottery draw or a mysterious date, we don't want to believe that it was just dumb luck. We want to believe that there is some order, some reason, some pattern to it all. We want to understand and take charge of our own fate. And why not—who wouldn't want to have power or knowledge over all of the random events that bombard us every day?

Indeed, many people believe that external items and events influence our luck. Rabbits' feet are supposed to bring good luck—originally, fertility luck because of rabbits' prolific breeding.[5] Four-leaf clovers, perhaps because they are so rare (they account for about one clover in 10,000), are also considered to be very lucky[6]—some even claim that Eve carried a four-leaf clover when expelled from Paradise. Horseshoes hanging in a house are said to bring good luck and protect from evil, perhaps due to a legend about a blacksmith using one to ward off the devil.[7] Walking under a ladder is generally considered unlucky, perhaps because of ladders' association with gallows (though this one at least has a practical side: if you walk under a ladder, someone might drop something on you).[8] Knocking on (or touching) wood is believed to ward off bad fortune, perhaps due to a pagan belief about gods who live in trees.[9] Crossing your fingers apparently originated from an ancient practice of marking a place for good spirits to concentrate.[10] And spilling salt is bad luck (perhaps because Judas supposedly spilled some at the Last Supper[11]), but it can be safely counteracted by tossing a pinch of the stuff over your left shoulder.

Meanwhile, birds have long been held to have special divining powers.[12] Today, this is reflected by snapping a turkey's wishbone, supposedly giving good fortune to whoever obtains the bigger piece (thus giving rise to the expression "lucky break"). An albatross, in particular, is considered to be good luck if it follows you, but bad luck if you kill it (as dramatized in Samuel Taylor Coleridge's famous poem "The Rime of the Ancient Mariner"). Jade jewellery is believed to bring wealth and friends.[13] Meanwhile, black cats' dark colour and mysterious nocturnal nature make them bad luck in the US—but, interestingly, good luck in the UK and Japan.[14] It gets a little hard to keep track. The one certainty is that lots of people believe in lots of different ways of mystically and magically summoning both good and bad luck.

But is it really true?

I have always been struck by the Serenity Prayer,[15] which asks God to "grant me the serenity to accept the things I cannot change, the courage to change the things I can, and the wisdom to know the difference." I don't always manage to live by those words, but at my best, I try to. After all, many aspects of our lives can and should be changed, but railing against things we cannot affect is a waste of time and effort and emotional energy.

I think a similar saying should apply to luck. Something like "Grant me the serenity to accept the random luck I cannot control, the knowledge to change the luck which can be modified, and the wisdom to know the difference." If we can figure out which lucky events are just random, dumb luck and which are caused by actual scientific influences—which ones can be affected and which cannot—then we can make better decisions, take more reasonable actions, and better understand the world around us.

I hope that, by the time you finish reading this book, you will have that luck serenity, that luck knowledge, and that luck wisdom.

CHAPTER 3

The Power of Luck

I t is all well and good to have stories about passport curses, lost jewellery, gambling odds, and dating hopes. But they all seem a bit light and fluffy. Is that all luck is? Does luck only affect the minutiae of our day-to-day lives? Are the weighty, serious, long-term outcomes all due to careful control, and scientific forces, and logical causes, and moral imperatives, involving fundamental meaning with little or no random luck at all?

Hardly.

As confusing as luck is—whatever luck means, however we interpret, explain, and justify it—it has a huge impact on our lives in all sorts of ways, big and small. It can reunite long-lost relatives and change people's lives. It can reveal hidden treasures that transform simple farmers into millionaires. And, tragically, it can even lead to the deaths of huge numbers of innocent people.

Hawaii Surprise

When Joe Parker offered to take a photo of a family at a beach in Hawaii,[1] he got more than he bargained for. He recognized the family's Massachusetts accent and, having grown up in central Massachusetts himself, he asked them where they were from and who they knew there. At one point, he asked if they knew someone named Dickie Halligan. The man replied that Halligan was his father. Astonished, Parker replied that Halligan was his father too.

A few minutes of chatting cleared up the confusion.[2] The man on the beach was named Rick Hill. And although Hill had never met Parker before, they were both sons of the late Dickie Halligan, born years apart to different mothers and growing up in separate, adjacent Massachusetts towns (Lunenburg and Leominster). In short, they were half-brothers, meeting for the first time as full-grown adults on a Hawaii beach.

The two men soon became fast friends and spent lots of time together. This was especially important for Parker, who had grown up in foster homes with little support and suddenly found himself welcomed into a stable family environment. His life was changed forever, and very much for the better, by a chance meeting on a beach. As Hill put it, "Joe found family with us, and my kids love him."

The story contains a few further twists. The Hill family hadn't planned to go to Waikiki that day, but decided on the spur of the moment to drop by. Parker wasn't supposed to be there, either, but went to secure a last-minute surfing lesson for a guest at the resort where he worked. As an added bonus, Parker celebrated his 38th birthday just six days later, in the company of his new family.

When I was interviewed about this story on television (for a program called *Supernatural Investigator*, whose name indicates the producers' perspective), I had to consider some important questions: Was Parker a lucky man? Had "fate" intervened to help turn his life around? Was it "destiny" that these two half-brothers should be united at last? Does this story show, once and for all, that luck and chance are imbued with special significance and are shaped by mysterious supernatural forces that help guide us to success and fulfillment? How should we decide?

A Special Co-Worker

Michigan resident Steve Flaig knew he was adopted, and he dreamt of one day reuniting with his birth mother. When he reached the age of 18, he enquired with the adoption agency that had placed him and was eventually given his mother's name: Christine Tallady. However, further enquiries went nowhere. So Flaig gave up his quest, carried on with his life, and got a job driving a delivery truck in Grand Rapids.

Four years later, Flaig casually mentioned this situation to his boss, including his mother's name. The boss casually asked, "You mean Chris Tallady, who works here?"

He referred Flaig to a woman who staffed the cash register in the very same store. Flaig had greeted that woman a few times, but had never learned her name. Now he knew. Further investigation confirmed that, incredibly, the woman at the front of the store was indeed Flaig's long-lost birth mother. After an awkward initial meeting, the two spoke for two and a half hours and became fast friends. After that, they regularly hugged each other whenever their shifts overlapped at the store where, *luckily*, they both worked.[3]

If Flaig had gotten a job driving a delivery truck at a different store, or if he hadn't bothered to mention his quest to his boss, or if his mother had found another cash register to staff instead, then he probably would have remained separated from his birth mother for the rest of his life. Their dramatic reunion was very lucky indeed.

So, was it just random luck? Or were other forces at work?

The Golden Farm

Eric Lawes was a retired farmer living a quiet life in the small village of Hoxne in eastern England. On November 16, 1992, his neighbour asked Lawes to help him find an old lump hammer that had been misplaced in his field. *Luckily*, Lawes was also an amateur metal detectorist, so he kindly picked up his equipment, followed his neighbour back to the field, and set to work.

He didn't find the hammer. Instead, he detected a pile of metal inside a decayed, buried wooden chest. Further investigation revealed that this was no scrap metal. Rather, it consisted of hundreds of gold and silver coins, jewellery, spoons, and other precious items of ancient Roman origin, dating from the fifth century, and worth over $4 million![4]

It was certainly lucky that the neighbour lost his hammer, lucky that he asked Lawes for help, lucky that Lawes knew about metal detection, and lucky that Lawes didn't simply find the hammer and stop looking. Most of all, it was incredibly lucky that Lawes instead stumbled upon this incredible gold and silver find.

Authorities were quickly called, and archaeologists excavated the site the following day. They classified a total of 7.7 pounds of gold, including 569 gold coins, and 52.4 pounds of silver, in what became known as the Hoxne Hoard. The items were all donated to the British Museum, where they remain on display. Lawes was paid a handsome £1.75 million finder's fee (which he kindly shared with his neighbour). The find is considered to be of great historical and archaeological importance.

Oh, and in the ensuing complete excavation, the neighbour's hammer was found too.

The Luck of Kokura

Luck doesn't just bring joy, like long-lost relatives and hidden treasures. At its worst, it can bring terrible, horrible death to many thousands of people.

On August 9, 1945, at 10:44 a.m., the United States B-29 bomber Bockscar flew over the Japanese city of Kokura, carrying the world's third-ever atomic bomb, Fat Man.[5] Three days earlier, a different atomic bomb, Little Boy, had been dropped on Hiroshima, killing 80,000 people instantly and about twice that number in the months following. Three weeks before that, the world's first-ever atom bomb,

Trinity, had been tested in the New Mexico desert. And now, on August 9, in a continuing effort to force Japan to surrender and thus end World War II, Kokura was to be the next target.

But something was wrong. Bockscar's arrival had been delayed due to a missing support plane, and by the time it finally got to Kokura, clouds had moved in. Despite three attempted bomb runs, the pilot could not see well enough to drop Fat Man at the desired spot. Finally, with fuel running low and enemy planes arriving, the bombing of Kokura was called off. The pilot instead flew off to the nearest secondary target, the city of Nagasaki, some 95 miles to the south. Fat Man was instead detonated there, killing at least 40,000 people instantly and about as many more in the months following.

What can we make of such a story? The skies over Kokura had been clear earlier that morning. Something as innocuous as an approaching cloud formation had saved literally tens of thousands of Kokurans from instant death, while killing tens of thousands of Nagasakians instead. Was this part of some grand plan? Did the Nagasakians "deserve" to die, while the Kokurans deserved to be saved? Was it fate that some died while others lived? Was there a reason for it all?

Or was the whole thing just the result of plain, dumb, horrible luck, with no logic or reason or explanation at all?

We humans have an instinctive desire to explain why things happen, to find a reason for them, to have them make some kind of sense. We rebel against unjustified outcomes. We want all of life's ups and downs to get tied up in a nice little bow. We want to believe that there was a young couple in Kokura, virtuous and kind and hopelessly in love, admired by all and hated by none, whose upcoming wedding just *had* to be saved by diverting the bomb elsewhere. Or, that there was an evil old man in Nagasaki, full of cruel plans to mistreat colleagues, harm strangers, and torture innocents, who just *had* to be killed off in the name of justice.

If this were a novel, or a Hollywood movie, then that would probably be the case. But in reality, out of tens of thousands of people, surely both Kokura and Nagasaki were home to many young couples and kind citizens who deserved to be saved. And surely both cities were also home to some cruel and evil people too. So what "reason" or "justice" could possibly be involved in an atomic bomb falling on one city and not the other?

None, it seems. Rather, on that day back in 1945, the people of Nagasaki were very, very, very unlucky. By contrast, in a twisted way, the people of Kokura were very lucky—tens of thousands of lives there were saved by a few clouds. Indeed, in Japan, the phrase "the luck of the Kokura" is still used to refer to this.

So perhaps luck isn't fair after all. Perhaps there isn't any deeper reason or meaning to the randomness we encounter. Perhaps tens of thousands of people die because of a few clouds, or suffer other terrible injury or death, just by horrific dumb luck, with no guiding principle at all. Is that all there is?

But wait. The atomic bomb that destroyed Nagasaki was built by human scientists. The choice to use it was made by human politicians. The plane that dropped it was flown by a human pilot. The decision to change targets was made by human generals. So perhaps humans are the only problem here. Perhaps humans, blessed with free will, do all sorts of things for all sorts of reasons, good and bad. Perhaps it is nature itself, and nature alone, that is governed by mysterious forces that guarantee fate and destiny and principles of fairness and justice.

Hardly. Deadly, destructive, unfair luck abounds in nature too.

A Random Tsunami

On December 26, 2004, one of the largest and longest-lasting earthquakes ever recorded occurred in the Indian Ocean, west of Indonesia. The resulting shock triggered a series of giant ocean waves, or tsunamis, to race towards distant shores in all directions. In the hours following, water crashed violently onto beaches around the world, killing at least 230,000 people in 14 different countries, displacing over a million people from their coastal homes and causing countless billions of dollars' worth of damages.[6]

So, was there some "reason" for this destruction? The earthquake began with the jostling of two of the Earth's tectonic plates. The Indian Plate slid about 15 metres under the Burma Plate, along a surface about 1,600 kilometres long. This slippage wasn't caused by any decision or action of any human. It just happened.

And what about all the people who died? Their only crime was being on or near an affected beach on the wrong day at the wrong time. Some of them were poor people living in primitive coastal huts. Others were rich tourists spending the day at a foreign beach resort. Surely among the hundreds of thousands of victims were some cruel evil-doers who might have deserved punishment. But just as surely the victims included plenty of good and honourable people who had worked hard and treated others fairly throughout their lives, only to see it all end with the coming wave.

So, did they deserve to die? Was death their fate? Their karma? Did these deaths have meaning? Were they just?

Jackpots and Cancer and Profits and Love

Luck doesn't end with family reunions and gold coins, nor with Kokura or the Indian Ocean. Indeed, every day, we are surrounded by events outside of our control. We get stuck in traffic. We run into a dear old friend. We get soaked in the rain. The train is running late. Our child aces an exam. The boss is in a bad mood. Our stocks go up. We catch a cold. We win a prize in a draw. Our washing machine breaks down. We buy a lottery ticket. We place a bet on a roulette wheel. In each case, our future well-being is affected by things controlled by other people (traffic, work, stock sales), by a physical mechanism (lottery draw, appliance design, roulette spin), or by nature itself (weather, germs). If these random elements work to our advantage, we call it good luck and bask in the rewards. If they work against us, we call it bad luck and complain bitterly. Either way, we feel a sense of frustration and mystery that we do not control our own destiny.

Meanwhile, each year, millions of innocent people around the world die of cancer and other horrible diseases. (Tragically, while I was writing this section, singing sensation Michael Bublé announced that his three-year-old son had been diagnosed with cancer,[7] causing him to put his career on hold—though fortunately the boy now appears to be recovering.[8]) People get injured, careers get ruined, good people suffer through no fault of their own. And elsewhere, someone wins a lottery jackpot and gets suddenly rich, while a colleague gets promoted simply by being in the right place at the right time, and a neighbour achieves great success purely through happenstance.

And so it has always been. Reflecting on the days of the Wild West, Sergio Leone (director of the classic western movie *The Good, the Bad and the Ugly*) is reported to have said, "In pursuit of profit there is no such thing as good and evil, generosity or deviousness; everything depends on chance, and not the best wins but the luckiest."[9] Indeed.

Oh, and when Clint Eastwood's character in *Dirty Harry* confronts a villain who is considering reaching for his rifle, he offers the following advice: "You've got to ask yourself one question: 'Do I feel lucky?'" Which, it seems, regardless of the time period, location, or situation, is always the most important question of them all.

Sometimes, luck gives and takes away with astonishing speed. Such was the case of Donald Savastano. On January 3, 2018, he won $1 million in New York State's Merry Millionaire lottery.[10] With his winnings he purchased, among other things, a medical checkup, which he couldn't afford before. Sadly the doctors discovered that he had stage 4 cancer in his brain and lungs. Mr. Savastano died on January 26, 2018, a mere 23 days after becoming a millionaire. You can't take it with you, indeed.

Even entire wars can be affected by luck. When Hitler's armies attacked Stalingrad in August of 1942, they expected a quick victory. However, *luckily* for the Russians, the winter of 1942–43 turned out to be extra cold.[11] This weakened the German army and ultimately contributed to its defeat in the brutal five-month-long Battle of Stalingrad. Without that luck, the Germans might have prevailed at Stalingrad. If so, they might have overwhelmed and conquered Russia. And if they did, if that winter had not been quite as cold, then World War II— and much of 20th-century European history—might have turned out very, very differently.

We first learn about luck at a very young age. Before babies can talk, they have little control over their destiny. Whether they are given their favourite toy, or have to settle for lesser entertainment, is entirely outside their control—that is, it's a matter of luck. And as they grow, every teenager knows the ups and downs of learning about romantic relationships. It seems quite difficult to predict or affect the extent to which you will or will not meet an appropriate partner and engage in a romantic connection. It's no surprise that success in this matter is called "getting lucky."

Not everyone is pleased by the influence of luck on our lives. In the classic novel *The Catcher in the Rye*, protagonist Holden Caulfield hears his teacher call to him, and reflects, "I'm pretty sure he yelled 'Good luck!' at me. I hope not. I hope to hell not. I'd never yell 'Good luck!' at anybody. It sounds terrible, when you think about it." Sometimes, even positive luck isn't pleasing. Sometimes, we would prefer not to be controlled by luck at all.

In a lovely commencement speech at his son's graduation, the chief justice of the US Supreme Court, John Roberts, said, "Commencement speakers will typically also wish you good luck and extend good wishes to you. I will not do that. . . . I wish you bad luck, again, from time to time so that you will be conscious of the role of chance in life and understand that your success is not completely deserved and that the failure of others is not completely deserved either."[12] An interesting perspective, that it is especially the *bad* luck that reminds us of the role of chance in our lives. But whether we are reminded or not, the chance and luck and randomness are always there.

It seems that so much of our lives is decided not by anything we do, nor even by our skills and abilities and who we are, but rather by surprises completely outside of our control. By, well, nothing but dumb luck. Could it be?

Buffett's Lottery

Another perspective on luck was provided by the billionaire Warren Buffett. Despite his wealth, he believed we should take care of the less fortunate. Why? Well, he imagined a scenario in which, on the day before we are born, we have to reach into a huge bucket containing one slip of paper for each human in the world. Whichever slip we draw determines who we are. As he put it, "You could be born intelligent or not intelligent, born healthy or disabled, born black or white, born in the US or in Bangladesh, etc." He argued that this scenario would make us want a world that treated everyone of every gender and race and background fairly—because that person could turn out to be us.[13]

And really, wasn't he onto something? As the saying goes, "There but for the grace of God go I." Indeed, for any of us to even be here, our parents had to meet and have a baby at just the right moment. And their parents in turn had to do the same. And all our ancestors before them. The fact that any of us exists, and that we look the way we do and were born in the place we were, to our own family— never mind being of good health and sufficient wealth and adequate safety—is, well, nothing but one big incredible stroke of luck.

CHAPTER 4

The Day I Was Born

I was born on October 13, 1967, at Scarborough General Hospital in Scarborough, Ontario, Canada, a suburb of Toronto. For many years, I didn't think too much about the actual date. Then, upon becoming a busy professor, I wrote a computer program to keep track of my appointments and plans—an early version of what is now called agenda or calendar software.

While trying out my new computer program, I discovered that it worked not only for planning future dates, but also for revisiting dates in the past. So I decided to go back in time—in my computer program, at least—and add a few important events from my earlier life. One of the most important, surely, was the day I was born. So I guided my computer program back to the year 1967, found October 13, and proudly typed in, "BORN!!"

As I typed, I noticed something interesting. My computer program arranged its dates according to the day of the week, and October 13, 1967, was displayed under the heading Friday. And that is when I realized, for the first time, that I was actually born on *Friday the Thirteenth*.

There is a long history in Western culture of fearing Friday the Thirteenth—so much so that it even has a technical name derived from Greek: *paraskevidekatriaphobia*, also called *friggatriskaideka-phobia* after the Norse goddess Frigg. The origins are many and varied. In the Christian bible, Jesus was betrayed and killed by Judas Iscariot, one of 13 people at the Last Supper—and was then killed on a Friday. In the middle ages, the French king, Philip IV, ordered the arrest of the Knights Templar on October 13, 1307, which was indeed a Friday (according to the Julian calendar in use at the time)—precisely 660 years before my own birth. The Italian composer Gioachino Rossini died on November 13, 1868—another Friday. US president Franklin D. Roosevelt apparently refused to fly on the 13th of a month (though he died on Thursday the 12th—one day early?).[1] There is also a Norse myth about Loki being the 13th guest to arrive at a dinner party in memory of Balder the Beautiful, whom he had killed, at which point the entire Earth supposedly went dark.[2] It has also been argued that 13 is the first "unfamiliar" number, since our clock faces, calendar months, multiplication tables, inches, and egg cartons all stop at 12.[3]

In 1907, the American businessman Thomas Lawson published a novel called *Friday the Thirteenth*, in which a stockbroker picks that particular day to wreak havoc on the financial world.[4] Later that same year, a ship financed by and named after Lawson perished in a storm off the coast of England in the early morning hours of Saturday, December 14, 1907. But because of the time zones, in Boston, where Lawson was living, it was still Friday the Thirteenth.[5]

In modern culture, many people take this fear surprisingly seriously, refusing to fly or take other "risks" on any Friday that falls on the 13th day of a month. Many skyscrapers are constructed, bizarrely, with no 13th floor. (More precisely, the 13th floor is labelled as the 14th, apparently in the hope that people living on that floor won't realize it is actually the 13th.) The author Sholem Aleichem, whose stories formed the basis of the musical *Fiddler on the Roof*, was so afraid of the number 13 that he numbered his 13th manuscript pages as "12a" instead.[6] When Princess Margaret, younger sister of Queen

Elizabeth II, was born in Scotland in 1930, her birth registration was delayed by a few days so that her birth wouldn't be listed as the 13th in the local parish register.[7] And people are very quick to point out unlucky events related to the number 13, such as the explosion of an oxygen tank during the Apollo 13 moon mission on April 13, 1970 (though that date was a Monday, not a Friday). A motorcycle event on Lake Erie, involving over 100,000 bikers, celebrates their rebel nature by holding gatherings on, yes, every Friday the Thirteenth.[8]

Many people I know—including well-educated scientists and scholars—say they don't *really* fear the number 13. But if pressed enough, they will still admit to having a bit of trepidation. (By contrast, in China, the unluckiest number is not 13 but 4. And indeed, many of my Chinese friends admit to a bit of trepidation about *that* number.)

Luckily, I had grown up in a fairly scientifically minded family, so I did not believe such superstitions. So when I realized I was born on Friday the Thirteenth, I was more amused than fearful. In fact, I took it as a badge of honour, using my career success to "prove" to others that Friday the Thirteenth wasn't so unlucky after all.

Did I ever worry that, by flaunting this superstition so brazenly, I was in fact inviting bad luck to rain upon me? No. Of course not. Certainly not. Never. Well, hardly ever, anyway. Well, occasionally, I guess. (I'm only human, after all.) But not usually.

On a more practical level, it did occur to me then—and I am reminded of it again while writing this book—that if something strangely bad ever *does* happen to me, like being killed by a freak bolt of lightning or a sudden mudslide, then some people will gleefully use my misfortune to prove that Friday the Thirteenth is unlucky after all. (Oh well, at least that might increase sales of this book.)

Of course, many famous people were born on Friday the Thirteenth, including the Oscar-winning actor Christopher Plummer, and Julia Louis-Dreyfus (who played Elaine on *Seinfeld*), and the musician Feist, and even Fidel Castro. The singer Taylor Swift was actually born on a *Wednesday* the Thirteenth, but had her 13th birthday on a Friday the Thirteenth; she considered this to be so lucky that she chose the Twitter handle @taylorswift13 and has been known to paint the number 13 on her hand before performing.[9] In fact, in the 1880s a "Thirteen Club" was formed to confront super-stition by meeting on the 13th of each month and dining 13 to a table; they then happily reported on their members' subsequent good health and fortune.[10]

However, all sorts of superstitious people find some reason to avoid the number 13. Well, not quite all. One self-proclaimed num-erologist solemnly declared about Friday the Thirteenth that "I am inclined to consider it a lucky day for some, an unlucky day for others and an average day for the rest of us."[11] Which, in fact, is just another way of saying that it has no effect whatsoever.

By the way, while I was in the midst of writing this book, I went on a lovely lecture tour in Australia. In each of the last two cities I visited, it so happened *just by luck* that my assigned hotel room was on the 13th floor, which they hadn't eliminated. In a strange way, I was delighted.

Attack of the 13s

One claim made about Friday the Thirteenth is that it comes up more often than other dates. This claim is used to demonstrate the power of Friday the Thirteenth and how we should fear it. But is it true?

Well, sort of.

Each year has 12 months, and each month has one day numbered 13. So there are 12 different 13th days per year, each of which *could* land on a Friday. There are seven days in a week, so all things being equal, Friday the Thirteenth should occur 12 / 7, or 1.7, times a year on average.

But how many times does Friday the Thirteenth really occur? Well, it depends on the year. For example, the years 2018, 2019, and

2020 each have two. But 2021 and 2022 have just one each. And 2026 has three.

What happens over the long term? Well, because 365 (the number of days in a year) doesn't divide evenly by seven, years cycle around between those beginning on Monday, those beginning on Tuesday, and so on. Also, every fourth year is a *leap year*. Putting this all together, we follow a regular, repeating 28-year cycle. For example, the year 2000 was a leap year beginning on a Saturday. So the year 2028 will again be a leap year beginning on a Saturday, just like the year 2000. Meanwhile, the years in between will follow their own patterns. The year 2001 was a non-leap year beginning on a Monday, and so 2029 will be too.

If we look at these 28-year cycles, we find that each contains a total of 48 days that are Friday the Thirteenth, which makes sense since $28 \times 12 / 7 = 48$. (For that matter, each 28-year cycle also contains 48 days that are Friday the 12th. Or Wednesday the 11th. Or Monday the 23rd. Spread out over each 28-year cycle, all the days balance out perfectly.)

Disappointed? Don't be. There is one more twist. We don't actually have a leap year *every* four years. To coordinate the progression of days (caused by the Earth spinning on its axis) with the progression of years (caused by the Earth orbiting around the sun), a funny rule was put in place. Yes, there is a leap year once every four years. *Unless* that year is a multiple of 100, like the year 2100, which will *not* be a leap year. Oh, unless the year is also a multiple of 400, like the year 2000 and the year 2400, which still *are* leap years. Confused yet? The bottom line is that years proceed pretty much on a 28-year cycle, except for a few twists and turns that mean their real progression is actually a 400-year cycle.

A 400-year cycle? It's true. Remember how the year 2000 was a leap year beginning on Saturday? Well, the year 2400—that's right, 400 years later—will again be a leap year beginning on Saturday. And that is the *true* cycle of the years.

Now, if we divide up the years into 400-year cycles, then how many Friday the Thirteenths will we have? It turns out we will have 688 of them. Is that a lot?

Well, if all days were equally likely, we would expect that, every 400 years, we would see 400 × 12 / 7, or about 685.7 days that were Friday the Thirteenth. So, 688 is *slightly more* than we would expect on average. By contrast, in every 400-year cycle, there are only 684 days that are Friday the 12th. Indeed, no other date falls on a Friday more often than the 13th, which is tied with the 6th, the 20th, and the 27th for the top spot, with 688 each. Even more dramatically, the 13th falls on Friday more often than any other day of the week—with Wednesday and Sunday each tying for second place with 687. *Phew!*

In short, considering these 400-year cycles, the days do *not* quite balance out exactly. Rather, Friday the Thirteenth has a slight edge on most other days. But only very slight—all other days occur nearly as often. So, does this make Friday the Thirteenth seem more powerful and intimidating to you?

Unlucky Day?

Of course, the real question about Friday the Thirteenth is: Is it really an unlucky day?

One study published in the *British Medical Journal* in 1993 indicated that, on Friday the Thirteenth, there were 1.4 percent fewer cars on the highway, but 44 percent more hospital admissions due to transportation accidents. However, the authors admit that the number of accidents they studied was "too small to allow meaningful analysis" and also wisely note that "those recording accident data may be more likely to record accidents on Friday the 13th," so the counts might be biased.[12]

Most strangely, one of the study's authors later stated that the study was actually a *joke*, writing that "the paper was just a bit of fun and not to be taken seriously," was "written with tongue firmly in cheek," and was "written for the Christmas edition of the *British Medical Journal*, which usually carries fun or spoof articles."[13] Indeed, to my surprise, the staid *British Medical Journal* apparently has something of a Christmastime tradition of encouraging "light-hearted fare and satire" involving "quirky research questions" and "a good dose of humour and entertainment."[14] However, the study apparently did use real data and true statistical analysis, so the unusual motivations of the authors are probably irrelevant.

A few other studies followed. One, in 2002, analyzed all traffic accident deaths in Finland during the 43 Friday the Thirteenths and 1,339 other Fridays between the years 1971 and 1997. They found that 1.03 Finnish men per million died in traffic accidents on an average Friday the Thirteenth, compared to 0.98 on other Fridays—hardly any difference at all. But for women, they found that an average of 0.47 per million died on Friday the Thirteenth, compared to just 0.29 on other Fridays, an increase of more than 60 percent, which is quite large and is statistically significant. The author speculated that, perhaps, "on Friday the 13th women who are susceptible to superstitions obsess that something unfortunate is going to happen, which causes anxiety and [. . .] could produce driving errors with fatal consequences."[15] In other words, perhaps the very fear of Friday the Thirteenth in turn led to worse driving and more accidents. *Perhaps.*

Two years later, a follow-up study considered all Finnish road accidents (not just deaths). It found just *slightly* more accidents and deaths on Friday the Thirteenth than on Friday the 6th and Friday the 20th, for men and women both, but not enough to show a statistically significant difference. They concluded, "This study could not find any indication of overrepresentation of women in injury crashes on Friday the 13th," and further declared that there was "no significant difference in injury accidents [. . .] among [the] three Fridays."[16]

Then, in 2008, the Dutch Centre for Insurance Statistics reported that Dutch insurers received an average of 7,800 traffic accident reports per Friday, but just 7,500 reports per Friday the Thirteenth, suggesting that perhaps Friday the Thirteenth was actually *safer* than the average Friday.[17] (However, at least one commentator noted that one reason for this slight decrease might be that Friday the Thirteenth automatically excludes certain holidays, like Christmas and New Year's Day, when traffic accidents may be more likely.)[18]

Meanwhile, some Swiss doctors analyzed hospital emergency room admissions in Bern in the year 2000. They found an average of 61.2 admissions per day, of which 20.6 were considered medical emergencies. For the specific date of October 13, 2000 (my 33rd birthday!), which was not only a Friday the Thirteenth but also featured a full moon, they found 62 admissions, of which 20 were medical emergencies—practically identical to the overall average. They rightly concluded from these numbers that "it seems that the rare constellation of Friday the 13th and full moon is not associated with a higher risk for injury and sickness, at least in the Swiss capital."[19]

In 2005, the British newspaper the *Telegraph* published an article entitled "Coincidence? 13 really is the unlucky number," which noted that, during the first 11 years of the UK National Lottery, the number 13 ball was drawn the fewest times of any: just 120 times. (By contrast, ball number 38 was drawn the most often: 182 times.)[20] Did this trend continue? At first, it did. For the National Lottery between November 1994 and October 2015, the number 13 was tied with 20 for the fewest draws (215 each).[21] But in October 2015, the lottery increased the number of balls available to be drawn, from 49 to 59, and since then the number 13 has been drawn 18 times, which places it right in the middle, nearly equal to the overall average of 19.2.[22]

What about other lottery draws? Well, during the two years from September 2015 to August 2017, the Powerball lottery's main (white) balls came up 13 a total of 11 times, slightly less than the overall average of 13.84.[23] And, over a 35-year period, Canada's Lotto 6/49 drew the number 13 ball a total of 402 times, a bit less frequently than the overall average of 428.3.[24] So I have to admit that the number 13 *does* have a bit of catching up to do. But not too much—these differences are very slight and are not statistically significant and are well within the expected range of random values[25] (for example, in Powerball, ball number 35 came up far less often than 13 did). That is, these lottery-ball counts could well arise from just random luck alone, and don't really signify anything about the number 13 at all.

CHAPTER 5

Our Love of Magic

These different stories about luck and its interpretations go round and round in my mind. In each case, I find myself considering other people's reactions to the events and contrasting them with my own.

It seems absurd to me that quoting from *Macbeth* could somehow cause misfortune, and yet that woman at the party was sure of it. It seems unlikely to me that a divine being would first hide a precious diamond and then reveal it, but my friend was sure that's what happened. There's no way that "karma" could cause craps dice to land differently, yet a prominent radio host had no doubts at all. I can't imagine that ESP actually helps when searching for lost bracelets, even if my audience member is sure it did. Nor that having the same car as a potential mate could forecast romantic success.

What special force could possibly draw a man to a beach to meet his half-brother? What special "destiny" would possibly decide to spare Kokura while condemning Nagasaki? How could a "system" of spaces on a lottery card possibly increase your chance of winning? How could hundreds of thousands of people all deserve to die, simply because they were near the water when a tsunami struck? Why would one particular number or day be any less lucky than any other?

Most important of all, what drives so many other people to believe in all of these magical forces, reasons, and destinies that I just can't see? Why don't they just accept the simple, straightforward explanation that misfortune and lost diamonds and happy romances and horrible tsunamis all happen due to dumb, meaningless luck, based on thousands of tiny, subtle scientific physical causes but no overarching morality, destiny, fate, or meaning? Why do they prefer to jump to alternative explanations involving karma and destiny and fate and magic?

At some point, it hit me. These straightforward explanations of dumb, meaningless luck are, well, *boring*. They have no excitement and mystery. They provide no intrigue or amusement. They offer no satisfaction or purpose. They give us no sense of meaning or importance. By contrast, karma and destiny and fate and magic feel way more meaningful and important. And it seems that, when people consider the luck and randomness around them, they *want* it to have a special significance or meaning. They crave magic.

Humans are *mostly* content with the real, everyday world of science and logic and cause and effect. A world where good and bad things happen, not necessarily because they should, but just by random luck. The reality that the good guys don't always win, love doesn't always conquer all, and not every success or failure has a moral justification. However, we feel a special excitement at the thought of events outside of this scientific world view, of magic and prophecies and supernatural occurrences in all their many forms, of strange and mysterious superstitions (one book lists over 500[1]), of secret forces that magically protect the innocent and punish the guilty, a place where everything happens for a reason, with special meanings and significance beyond the apparent randomness of our daily lives.

And nowhere is this clearer than in works of fiction.

Magical Fiction

M. Night Shyamalan, the director of the movie *The Sixth Sense*, once said, "In all great movies—you know, the great, great, great movies—there's some element of magic, you know?"[2] Now, he might have been exaggerating slightly. After all, Humphrey Bogart's *Casablanca* is surely one of the greatest movies ever, yet it is completely reality based, without—so far as I can detect—any hint of magic or supernatural elements at all. (In fact, it was made in 1942, while the Second World War raged, making it all the more real.)

But Shyamalan's wider point, that elements of magic lend excitement and intrigue and appeal to many movies and other fictional works, seems indisputable. What sorts of magic? In Shyamalan's case, it was the ability of a young boy, Cole, to converse with dead people. In *The Wizard of Oz*, it is Dorothy returning to Kansas by clicking the heels of her ruby-red slippers. In *Heaven Can Wait*, it is Warren Beatty magically reuniting with Julie Christie after he has been transformed and lost his memory of her.

What do these magic movie moments have in common? They all happen for a *reason*. Cole receives information from the dead that allows him to save a girl's life and to reassure his mother that her own mother was proud of her. Dorothy manages to escape from Oz and return to the family that loves her. Beatty and Christie get to fall in love all over again. Fiction is replete with luck, of a sort. But it is never dumb, random luck. The luck in fiction always has meaning—and magic. Indeed, in fiction, if an event is boring and meaningless, it will get cut from the script. The only kind of luck that fiction permits is the interesting, exciting, meaningful, magical kind.

In the recent Hollywood blockbuster *Dr. Strange*, starring Benedict Cumberbatch, the title character injures his hands in a car accident, to the point where he can no longer perform the brilliant neurosurgeries that made him so successful. In the real world, that would be the end of the story. He would be forced to switch careers and live out his diminished life as best as he could. But not in fiction. Instead, after much searching, the doctor finds a sorcerer who can teach him special curative powers, explaining that she deals with "the source code that shapes reality. We harness energy, drawn from other dimensions of the Multiverse, to cast spells, to conjure shields and weapons. To make magic." There is that word again: magic. And did audiences appreciate this magical solution to Dr. Strange's woes? Well, let's just say that, in its first four months, the movie grossed US $677,718,395[3] and leave it at that.

One of my favourite books is *Shoeless Joe* by W.P. Kinsella, later the basis for the movie *Field of Dreams*, starring Kevin Costner. A corn farmer hears a mysterious voice say, "If you build it, he will come." Does he dismiss the voice as a meaningless figment of his imagination, or view it as a sign of mental illness? Does he point out that a mysterious voice could not possibly know what he should do, nor what will happen next? No. Instead, he knows right away that he has to, of all things, carve out a baseball diamond in the middle of his corn field. He also knows that this will, in turn, allow the legendary, long-dead baseball player Shoeless Joe Jackson to magically appear at his farm and play some friendly games of ball. After months of struggle, the farmer succeeds in building the diamond, Jackson arrives, the farmer's own deceased father eventually makes an appearance too, and the farmer experiences true and meaningful closure.

Could this happen in real life? Of course not. No mysterious voice can guide our choices, solve our problems, or bring dead baseball players back to life. In real life, it would be the craziest dumb luck if building a baseball diamond actually led to positive life changes. And if a man hears a strange voice in his head, it is much more likely to be a meaningless hallucination than a valuable instruction. But as fiction, the story works beautifully. The voice provides needed goals and focus, and the ensuing magic makes good things happen just when they are most needed. The ending is satisfying, touching, heartwarming, and meaningful on a deep and profound level in which the fictional magic plays an essential role. "If you build it, he will come," indeed.

I've mentioned that my favourite Shakespeare play is *Macbeth*. Much of the plot is driven by three witches, who tell Macbeth he "shalt be king hereafter." Does he follow the appropriate real-life course of action by dismissing them as a bunch of crazies whose words are meaningless? No, their words get him and his wife thinking—and, a few illicit murders later, he does indeed become king. The witches also declare that Macbeth's fellow nobleman, Banquo, "shalt get kings," meaning that his descendants (and not Macbeth's) shall be kings. Macbeth tries to thwart this prophecy by having Banquo murdered—but his son Fleance escapes, presumably to father future kings. The witches later warn Macbeth to "beware Macduff"—and rightly so, since Macduff will eventually kill him! But they also promise that "Macbeth shall never vanquished be until Great Birnam Wood to high Dunsinane Hill shall come against him," a development that seems impossible, and thus the prophecy seems quite comforting to Macbeth. Until, that is, Malcolm instructs, "Let every soldier hew him down a bough and bear 't before him"—that is, each of his men should lug one piece of wood from the forest onwards to Dunsinane. And why should they do this? Why, in order to "shadow the numbers of our host and make discovery err in report of us"—that is, to hide the true number of soldiers in their army. A clever military strategy? Perhaps. But more importantly, a nifty way to make the witches' prophecy come true, as it somehow must.

Perhaps most dramatically, the witches also assure that "none of woman born shall harm Macbeth." This emboldens Macbeth and makes him feel invincible, so in his final battle of the swords, he tells his accuser Macduff to "let fall thy blade on vulnerable crests; I bear a charmed life, which must not yield to one of woman born." Unfortunately Macduff replies by explaining that he was actually "from his mother's womb untimely ripped"—delivered by a primitive sort of Caesarean section (probably after his mother was already dead). Apparently such a beginning does not count as "born," so Macduff is not limited by the witches' prophecy. A technicality? Sure. But also, a clever and unexpected manner of fulfilling another prophecy. In response, Macbeth can do no more than sputter, "Accursed be that tongue that tells me so," before falling to Macduff's sword—and suffering the indignity of having his severed head paraded around the stage, to boot.

In fact, in *Macbeth*, even the weather is magically prophetic. Upon Macbeth's first murder, it is reported that "the night has been unruly . . . Our chimneys were blown down . . . the earth was feverous and did shake," as if the weather itself could detect that evil was afoot. In response to this, Macbeth can do no more than sputter, "'Twas a rough night."

Why are these prophecies so compelling? In real life, they would just be meaningless words, having no effect on the upcoming random luck of life and death and battles and crownings. However, this being fiction, the prophecies just *have* to come true. If Macduff had managed to kill Macbeth *without* the "untimely ripp'd" explanation, or without Birnam Wood having moved, then the play would have been pretty stupid. We would leave the theatre muttering, "Uh, I guess those witches were wrong. Uh, maybe they weren't so smart after all. Oh well." We would have wondered: What was the point of the witches appearing in the play at all? Why did we waste our time listening to their wild ravings? Why did the author bother with such phony misdirection? The play wouldn't capture our attention in the same way. It wouldn't be as intriguing or fascinating or lively. It wouldn't be, well, *Shakespeare*.

Magical Belief

Who among us has never been to a "magic show" in which some magician dazzles us with seemingly impossible feats, each one designed to make us believe less in the world of facts and more in the world of mysterious magic? How do we react to such displays? Well, there are always some people who try to figure out how the trick was done, or look for hidden wires, or check the magician's sleeve, or try to fish around in his pocket. But, for the most part, we don't want to know how it's done. We don't want to have the trick "explained" and brought down to cold, hard reality. Rather, we want to believe that we have witnessed, well, magic.

A professional magician friend[4] kindly surveyed several of his performer colleagues on my behalf. While they disagreed on the percentages, they agreed nearly unanimously that at least half of their audiences either believe the magic is real or at least pretend that it is.

Our love of magic extends beyond traditional magic shows. On a recent flight, I leafed through the airline's travel magazine and came across an article by a New York–based travel writer about visiting the "haunted" Merchant's House Museum in Greenwich Village. She wonders if this house could really be haunted.

The writer admits to harbouring doubts. But then, interestingly, she adds, "But I want to believe: in ghosts, in spirits, in something other than New York's prohibitively high cost of living."[5] Which nicely summarizes our attraction to supernatural magic in a nutshell. Faced with drab real-world circumstances like expensive rent, we would rather escape into the exciting, unknown world of ghosts who don't bother with silly, practical matters like paying rent. Indeed, while I was revising this chapter, it emerged that a controversial judicial nominee in the United States had spent much time investigating and writing about, yes, haunted houses. It seems there is no escape.[6]

Magic Dice

Many board games involve randomness: rolling dice, dealing cards, spinning wheels, picking tiles from a bag, and so on. Why? Because it's fun! Just as randomness adds excitement to games in a casino, it also adds an extra spark of mystery and intrigue to many board games. Such popular games as backgammon, Scrabble, Monopoly, Clue, Risk, bridge, and Settlers of Catan[7] just couldn't be played without the randomness of cards, dice, or tiles. Compared to non-random games like chess and checkers, they are more amusing, more dramatic, and often more enjoyable to play.

Even so, not everyone wants their random games to be truly random. They want the luck of games to be meaningful luck, something they can control and influence in mysterious ways. The other day, I found myself playing a complicated board game involving dice. To my surprise, each time my teammate rolled the dice, she first commented on what a big "responsibility" it was, and tried to figure out what numbers would be most advantageous to us, before actually rolling. What was the point of that?

I meekly pointed out that there was no need to figure out in advance which roll was best, since the result of the roll was random in any case. But she was quite resistant to this argument. She wanted to feel like she had some control of the dice, so she could put some effort into "trying" to get a good roll. Apparently this was much more fun and engaging and meaningful than simply accepting that the dice were entirely random, and there was nothing at all that she could do to change them (well, aside from cheating, which she had no interest in).

After several more turns, my teammate did gradually sort of accept my point of view. But she had the last laugh. On one turn, it seemed that a low roll would work best for us, but I rolled a pair of fives—not good. Upon seeing my result, she triumphantly declared, "I *knew* that would happen!" Sigh.

Curse of the Bambino

When the Boston Red Sox went 86 years without winning baseball's World Series championship, fans couldn't accept it as mere random luck or honest losses. Instead, they decided the blame lay with a magical "Curse of the Bambino," brought on by their team's terrible error in selling star Babe "Bambino" Ruth to the New York Yankees way back in 1919. Every heartbreaking loss was then further "proof" of the curse.

For example, in 2003, Boston was tied 5–5 with the Yankees in the 11th inning of the seventh and final game of the American League Championship Series (ALCS). The winner of this game would go on to play in the World Series. In the bottom of the 11th, recently acquired Yankees player Aaron Boone hit a home run to defeat the Red Sox.[8] Boston fans immediately declared that since Boone was a new Yankee, and the home run came in the 11th inning, it clearly demonstrated that the curse was at work. However, I'm pretty sure that if, instead, a long-standing Yankee had defeated them, or if the home run had come in the first inning, they surely would have declared that to be evidence of the curse too.

The next year, 2004, the Red Sox finally won the World Series—a victory made all the sweeter because they came back after losing the first three games of that year's ALCS to the Yankees. To me, this demonstrated that there hadn't been any curse after all. But to many Red Sox fans, it instead showed that the curse had finally been "broken."

There is a traffic sign on the Longfellow Bridge over Boston's Charles River, warning of a "reverse curve" in the road. At some point, a graffiti artist had modified this sign to instead say "Reverse the Curse." No one dared to clean off the graffiti, so it remained for years. Then, when the Red Sox finally won in 2004, the sign was helpfully edited to read "Curse Reversed."

It seems that when it comes to your favourite sports team, it is not enough to explain their wins and losses in such mundane terms as player acquisitions, physical abilities, strategic decisions, and coaching philosophies. Rather, their performance must be the result of something else. A curse. A superstition. Fate. Destiny. Something, well, just a little bit magical.

Oh, and what about when the Chicago Cubs finally won the World Series in 2016, after a 108-year drought? Was this just the end of a 108-year run of random bad luck? Of course not. It was immediately declared that they, too, had ended a magical curse—this one apparently instigated in 1945 when a goat was denied entry to a World Series game at Chicago's Wrigley Field.[9] (Why they didn't win between 1909 and 1944 was not explained.) Some things never change.

Lucky In Love

A recent web posting described a young woman who had had a crush on a friend since fifth grade. One day during high school, they agreed to meet. She explained, "After his mom dropped him off, we went on a walk by the stream. Then we sat on a bench. And I saw him flip a coin. So I thought he was bored. But right after, he gave me a kiss."[10] Lucky her! It seems that her friend was too uncertain to initiate a kiss on his own. Instead, he flipped a coin to decide whether or not to proceed. I guess he needed some random luck to work up his courage and provide him with a reason for proceeding. If they had been less lucky, and the coin had landed differently, then love might have been lost forever.

And why not? In a scene from the movie *Serendipity*, Kate Beckinsale's character meets John Cusack and decides that fate should decide their romantic future. She insists that they enter two separate elevators in a tall building and each choose a random floor. If they happen to select the same floor, she explains, that will show that they are fated to be together. He reluctantly agrees, and the experiment begins. Sure enough, they both make the same choice: the 23rd (and top) floor. This "proves" that they are meant to be together, right? Of course, in true movie tradition, harsh reality intervenes—this time, in the form of an obnoxious kid who gets on Cusack's elevator and pushes a bunch of buttons, thus delaying his arrival on the 23rd floor until after Beckinsale departs—and it takes the rest of the movie for them to finally reconnect. But in the end, fate cannot be stopped.[11]

It seems that not only is love a magical feeling, but we also want it to have special meaning by being controlled by magical, fateful, forceful luck.

Give Me a Sign

At a dinner party, a friend once told a story about a big old tree in her neighbourhood that was scheduled to be cut down to make way for a new building. My friend, who loved the tree, found this plan extremely upsetting. She explained that, the evening before the scheduled chopping, she visited the tree, inconsolable, bawling her eyes out. What could be done to save this wonderful tree from destruction? How could she make things right again?

In her desperation, she pleaded to the universe for help. "Give me a sign, some sort of a sign," she begged out loud. Just then, she looked up and saw the setting sun's rays shining brightly between two large branches of the tree, gloriously spreading light upon her. This pleased her greatly and allowed her to get through the difficult moment.

At the conclusion of her story, the other party guests nodded in agreement. They all understood her plight and appreciated the sign sent by the universe to comfort her. That made it all okay, they seemed to agree.

But not me. I was puzzled. The tree was still cut down on schedule, so my friend's "sign" didn't actually save anything. If the universe really was making a special effort to help my friend through supernatural means, wouldn't it have been more useful to actually save the tree than to merely shine some sunlight as a temporary, ineffectual signal? Furthermore, if the sun is low in the sky, and you are standing near a large tree, then you can always find some angle from which the sun is shining through some branches. What of it?

Scientifically speaking, I still think I was correct. There is no evidence that the universe was in any way intentionally conspiring to make the sun shine where it did, and indeed the occurrence was unremarkable and unsurprising, not to mention unhelpful for the tree.

But in a deeper sense, perhaps my friend was right after all. She had been very upset and distressed, and had found some way to feel better about it. The setting sun had, for her, injected an element of magic, of meaning, of harmony into a difficult situation. And who am I to argue with that?

Superstitious Me

Am I myself immune to superstitious, magical thoughts? I like to think I am. After all, I am a scientist. A rationalist. A skeptic. Someone who understands the nature of probability and randomness. Someone who realizes that real-life occurrences happen for many reasons and meaning is not guaranteed.

So, am I completely free of supernatural beliefs? Well, er, um, not entirely.

At my childhood birthdays, my mother told me that, before blowing out the candles on my birthday cake, I should make a wish, one that would come true provided I kept it a secret. I mostly just went along with this practice to be a good sport, without actually believing it. But perhaps a small part of me did believe, anyway. I remember once accidentally revealing some information about my wish to my mother, and then immediately feeling terrible. I feared that I had ruined my wish once and for all, so that it would never come true. Could it be? And, now that I am much older and wiser, do I still have such silly feelings? Not really. But maybe just a little, tiny bit.

When we ate turkey at Thanksgiving, my mother would carefully dry out the wishbone. Then, a few days later, she would let me and my brother pull at opposite ends. I still remember my disappointment one year, when my brother got the bigger piece, and I figured I was doomed to a year of misfortune. Even more so, I remember that fighting over the turkey bone, and believing in its special magic powers to mysteriously control a year of luck, was, well, pretty *fun*.

When I grew and learned to drive, I found that for me, like for many people, getting stuck in traffic brings out the worst instincts. There is a traffic light near my house, which I am convinced is red every time I approach it. No matter what time of day, or how late I am running, or how challenging my drive has been, that one traffic light is always red, and I always have to wait for it. Is this really true? Or do I just imagine that it is, and forget the times when it was green?

On a more serious note, my city of Toronto is mad about hockey. So, when was the last time the Toronto Maple Leafs won the big Stanley Cup championship? Oh, on May 2, 1967, five months and 11 days before I was born. That's right: my city hasn't won a hockey championship in my entire, long lifetime. Now, *of course* I know that my life hasn't had any effect on the hockey scores. Still, there is a small part of me that wonders, completely irrationally, if their next championship will come five months after I die. On the other hand, I sure hope that no Maple Leafs fans take that theory so seriously that they try to kill me!

My superstition doesn't end there. While I was in the middle of writing this chapter, Donald Trump was elected president of the United States. This was quite a shock to many people (including myself) who found him singularly unqualified for the office, and who also had seen numerous opinion polls suggesting his opponent, Hillary Clinton, had a small but significant lead.

The day before the election, a political scientist convinced me to place a small bet on the election: I would pay him one dollar if Clinton won, while he would pay me two dollars if Trump won. I was actually more than 2-to-1 confident that Clinton would win, but I took the bet anyway, just for fun. Then, when Trump won, I consoled myself that, although I wasn't happy with the election result, at least I had outsmarted my colleague and won two dollars.

Then the strangest thing happened. I started to get a weird, uncomfortable feeling. As I drifted off to sleep, I found myself fearing that perhaps Trump had won the election precisely *because* I'd bet that he would. Maybe I had brought the Trump presidency upon myself! Rationally speaking, there was no way that my one little bet, known to almost no one, could possibly have affected even one vote. But my uneasy feeling persisted until morning. It seemed that a part of me wanted to believe in some sort of magical luck by which my silly bet could somehow "explain" the shocking election result. Madness.

Meaningless Misery

Perhaps the best way to see the importance of meaning in fiction's randomness is to consider the rare occasions when it isn't there.

For example, the television series *Star Trek: The Next Generation* initially featured a feisty female security officer named Tasha Yar. Towards the end of the first season, Yar was suddenly killed by a bolt of energy from an evil alien slime creature. Fans were, of course, disappointed and shocked to see a beloved regular character killed off from their show. But something else was afoot. It quickly became clear that this particular death was very unsatisfying. Why? Because it was meaningless. Yar's death had served no purpose, advanced no story line, created no magic, achieved no goal. It just simply was. (Indeed, apparently the death was arranged because the actress who played Yar wished to leave the series.)

The dissatisfaction among fans was so great that action had to be taken. A complicated plot was arranged involving time travel and alternate realities, all in an effort to create some circumstance in which Yar could appear again. And sure enough, after many contortions, there she was, back on the small screen. So, what happened next in the episode? Yar died again! But this time, it was different: she died while guiding another ship to heroically battle the enemies and prevent an all-out war. Death remained, but meaning and magic were restored. In the end, everyone felt better and more satisfied.

By contrast, I once saw an Australian movie, at a film festival, in which a character is carrying a bunch of lottery scratch tickets and accidentally drops one without noticing. The camera shows a close-up of the ticket, so we all know it is important and has special meaning. Later, another character sees the ticket lying on the ground, picks it up, and scratches it. So, does he then start jumping up and down and celebrating his big lottery win? No, he just disappointedly casts it aside, winning nothing. Apparently the ticket was a dud after all.[12]

When I saw that movie, I felt cheated. Why would they emphasize this lost lottery ticket so much, only to have it turn out to mean nothing? What a waste of our time and attention. I was reminded of Anton Chekhov's famous dictum that "One must not put a loaded rifle on the stage if no one is thinking of firing it."[13] Indeed, if the great Russian playwright couldn't show a meaningless gun, then how could this Australian movie show a meaningless lottery ticket? It was wrong. It was unfair!

Only later did I realize that the movie actually had it right after all. In real life, as opposed to fiction, most things that happen do *not* have a special meaning or work out just so. And most lottery tickets do *not* provide a big jackpot. The Australian movie was simply showing things the way they really are. But that's not what audiences want to see.

CHAPTER 6

Sharpshooter Luck Traps

I t seems clear that in fiction and beyond, people want to find meaning and magic in their daily surprises. They are not content to accept strange occurrences as dumb, meaningless luck.

But what about in real life? When something dramatic or unexpected happens, how can we evaluate it? How can we determine whether it is meaningful or just random? How can we determine, as in the luck Serenity Prayer, which luck is controllable and significant, and which is not?

Often an outcome or story or anecdote will *seem* to be meaningful. But is it really? It turns out that there are lots of luck traps that can trick us into misunderstanding and misinterpreting the evidence before us.

To put this into concrete terms, suppose you want to find an expert sharpshooter, someone who can shoot a gun so accurately that they can represent your city in a big international competition. How would you go about it?

Suppose you set up a target and paint a line far away. You tell your assistant that when an applicant arrives, they should be positioned behind the line and told to shoot at the target. You figure that if they hit the target, they must be a good shot, so you will hire them.

A fine, logical test. So far, so good.

Suppose that later, your assistant comes running up to tell you that an applicant arrived and did indeed hit the target. Great! Then you should hire them, right? Well, maybe. But maybe not.

Possible Explanations

Before signing up this applicant, you decide to think everything over one last time. You decide to make a list of all the possible explanations for what has happened. It turns out that there are many.

True Skill. The happiest explanation—the one you hope is true—is that the applicant does indeed have *true skill*. They can indeed shoot well, and hit the target, and pass the test fairly, any time they want to. So you should choose them, no question. But what about other possible explanations?

Lucky Shot. At the other extreme, perhaps the applicant is terrible at shooting, but just made a really, really lucky shot and hit the target anyway, just this one time in a million. This is the alternative that statisticians consider most often, using fancy statistical tests. If the chance of a lucky shot is really, really small, then that probably isn't the true explanation, so we can probably eliminate it. Well, then, if it wasn't a lucky shot, then it must be true skill, right? Not necessarily— there are many other possible explanations.

Shotgun Effect. Perhaps the applicant used a shotgun instead of a regular gun. As a result, lead from their shot travelled all over the place. Yes, some of it hit the target, as your assistant saw. But most of it missed! This doesn't prove they're a good shot at all, even though they did succeed in hitting the target.

Many Tries. Similarly, perhaps the applicant fired at the target over and over again, dozens of times, until they finally hit the target. Does this prove they're a good shot? No, not at all—it's similar to the shot-gun effect. But if your assistant only tells you about the one good shot, you might think they are.

Many People. Perhaps your assistant actually tested a thousand different shooters, and this applicant was the only one who hit the target. If so, then this is also similar to the shotgun effect—given such a large number of applicants, one of them is bound to eventually hit the target by luck alone, even if none of them have any shooting skill whatsoever.

These last three possibilities are all similar. They all boil down to saying that even if an applicant hit the target, this successful effort may have been just one of many attempts, whether due to a spread of lead shot, the applicant firing lots of times, or lots of different people each firing once. I sometimes refer to this as the *out of how many* principle: it's not enough to know that someone succeeded once; the question is once out of how many different tries.

Some other possible explanations involve cases where hitting the target wasn't as hard as you thought:

Extra-Large Target. Perhaps someone replaced the target with a much, much larger one, and then hit the new target. This made their job much easier. In that case, hitting the target might not be so impressive, and might not indicate that the applicant is such a great shot after all.

Hidden Help. Perhaps the applicant didn't stay behind the line, but drifted (perhaps unintentionally) much closer to the target and shot from there instead. Or perhaps they had a special laser-guided rifle that always hits its target. Or perhaps the target contained a magnet that pulled the bullet towards it. If the applicant hit the target with extra help, this doesn't mean nearly as much—but you might think it does.

And we're not done yet. There are other possible luck traps too. For example:

False Reporting. Perhaps what your assistant told you is simply false, and the applicant never hit the target at all. Maybe the assistant is flat-out lying. Or perhaps he didn't remember quite right, or he got confused, or he misinterpreted a small tear as a bullet hole. If you didn't see the shot yourself, you can never be completely sure.

Placebo Effect. Maybe your assistant was swayed by *psychological factors*. For example, perhaps the candidate spoke with confidence, looked handsome and rugged, and bragged about his shooting abilities, to the point where the assistant thought the candidate had hit the target when he really hadn't.

Alternate Cause. Maybe there is some completely different reason why a bullet hit the target. For example, perhaps an expert marksman fired his own shot at the target, at the exact same moment as the applicant, and the expert marksman's bullet hit the target while the applicant's bullet went flying through the woods. In this case, the applicant's ability had no bearing at all and should be ignored.

The above three luck traps describe situations where the cause of the result is quite different from what it initially seems and does not actually provide any evidence at all in favour of the applicant.

And here's one last consideration:

Different Meaning. Even if the reporting is all true, and the applicant is indeed good at shooting at a target, it might not mean what you think it does. For example, perhaps the gun used for the test is completely different than the one that will be used in the actual competition. Or perhaps hitting a target under calm conditions is very different from being able to shoot well when the pressure is on. In this case, even if the test was accurate as far as it went, you still have to be careful about interpreting the results correctly.

To evaluate your sharpshooter applicant's abilities, you need to try to figure out which of these explanations is correct. Only once you have eliminated all of the other possible explanations—not just *lucky shot*, but *shotgun effect* and *many tries* and so on—can you safely conclude that your applicant does indeed have true skill and should be hired.

Multiple Evidence

The above luck traps all apply to evaluating just a single shot. There are a few other explanations that arise when considering a longer test involving many different bits of evidence.

For example, suppose your assistant says that the applicant hit the target five different times. You would probably be very impressed. Even if one shot could be written off as luck, surely five hits is indicative of an excellent shooter, right? Well, probably. But you had better watch out for:

Biased Observation. Suppose your assistant only told you about the few shots the applicant made, but didn't tell you about the thousand other shots that missed. Then you might think the applicant always (or usually) hits the target, when the reality is quite different. Your assistant's reporting was biased—you were only informed of the hits, not the misses—so you were only able to observe certain types of events, giving you a misleading impression.

Similarly, if the assistant only told you about the few shooters who managed to hit the target, without mentioning the dozens who failed, then you might think that everyone in town is a good shot, when in reality most of them are terrible.

So whatever fact you hear, you should check that you are getting the whole story—or at least a representative sample, not a biased sample that skews your impressions.

And there's more. Suppose your assistant comes running and tells you that he found a candidate who not only hit the target, but also owns his own gun, is a member of the local rifle club, has steady hands, and frequently practises his shooting. Any one of those attributes is good, he explains, but finding all of them in the same person is fantastic! Should you be especially impressed by this? Not really—it is an instance of:

Facts That Go Together. If someone owns a rifle, he is more likely to be a member of a rifle club. And if he is a member of a rifle club, he is quite likely to attend shooting practice. And if he practises often, he is probably a good shot who can hit the target. And if he likes shooting, it probably helps that he has steady hands. That is, if an applicant has one of these various attributes, it's not surprising that he has the rest of them too. In this case, we should still count his hitting the target as good evidence. But we shouldn't count as extra evidence—at least, not too much extra—the rifle club membership and gun ownership and steady hands and so on, since they tend to go together with being a good shot anyway.

This last issue is related to what I call the *to multiply or not to multiply* question. If you have two coins, and each has one chance in two of coming up heads, then the probability they will both be heads is $(1/2) \times (1/2) = 1/4$. Since neither coin affects the other, we get to multiply the probabilities together. But if one-tenth of the people in your town own a rifle and one-tenth belong to the rifle club, then the probability of people both owning a rifle and belonging to the rifle club isn't $(1/10) \times (1/10) = 1/100$; it is much greater—probably nearly $1/10$. Why? Because these two items tend to go together, so it is incorrect to multiply their probabilities.

The bottom line is that the more evidence you have, the more convinced you should be—usually. But if the evidence is biased, or all tends to go together, then you should proceed with caution, since the extra evidence may not be truly extra after all.

So, why do I tell you all about these potential luck traps that offer alternative explanations for the sharpshooter's success? After all, I don't care much about sharpshooters, and maybe you don't either.

The reason is simple: these same sorts of luck traps apply to so many other stories too. Indeed, I would say that virtually every other story about luck, from casual discussions to academic research, can also be thought of in terms of the sharpshooter's luck traps.

Yes, even those lucky tales that discomforted me so.

CHAPTER 7

Luck Revisited

can our sharpshooter luck traps provide new insights into tales of
luck? Yes, indeed. In every story we have discussed so far, and in
so many others, a careful consideration of luck traps can help to
reveal hidden truths and avoid false conclusions.

Let's reconsider those earlier tales of luck, one by one.

Half-Brother Half-Explanation

Those reunited half-brothers provide an interesting case study.
The story is so remarkable—meeting for the first time on a remote
Hawaiian beach, discovering their special connection, becoming
friends, changing their life directions—that it certainly feels like the
encounter has some underlying meaning. And indeed, the chance
that these two specific half-brothers would meet on this one particular day at this one exact beach is extremely unlikely, to put it mildly.

And there's no *false reporting* here—the story appears to be
completely true. Nor is there any *extra-large target*: this reunion
was very special and unique. Nor any *hidden help*: no one told the
half-brothers which beach to attend on which day. Nor any *different
meaning*: this story was indeed very touching and significant, just as
everyone thinks it is.

So, is there any luck trap here at all? Yes, indeed: *many tries*. There are over 300 million people in the United States. Many of them—surely at least a million—have at least one close relative that they are estranged from in some way, whom they would find it very meaningful and special to run into. And each of those people makes countless decisions every day about what to do and where to visit. Out of those million pairs of people, and all of the days of their lives, and all of their associated actions and movements, in this *one* case a very special reunion took place. With so many tries, such an amazing event could well happen just by chance alone. So although the story is incredibly special and meaningful for the people involved, it does not necessarily indicate anything about a grand design to the universe that magically forced this reunion to occur.

Or, to put it differently: If you really do believe it was fate, karma, God, or destiny that caused these estranged half-brothers to meet on that beach, then why didn't that same special force help all the *other* people who are estranged from their families but never run into them on a beach?

Scottish Rejoinder

And what about that partygoer who felt that bad luck always follows those who recite lines from *Macbeth*, and whose "proof" was that, a week after quoting *Macbeth*, her daughter discovered that her passport had expired?

Well, first of all, the inconvenience of renewing a passport is the sort of mild setback that we all experience frequently, whether we quote *Macbeth* or not. If I manage to get through a week with only one difficulty of that sort, I consider myself quite lucky indeed. So, an *extra-large target* was being used here, in which any minor inconvenience was counted as "bad luck."

Related to this is a *shotgun effect*. Over the course of a week, the daughter surely had many experiences that *could* have gone badly, but didn't. If her food had spoiled, or she had slipped on the ice, or she'd had a fight with her husband, or she'd lost a game of cards, any of those could have counted as bad luck too.

In addition, there was probably also some *biased observation* at play. Over the years, various other people must have quoted *Macbeth* in this same partygoer's presence, but didn't then report any follow-up misfortune. And many others must have experienced misfortune, even though *Macbeth* never entered their minds. Were these facts counted as evidence against the superstition? Or were those instances ignored or forgotten, with inconveniences only noticed when they happen to fit in with a certain preformed superstitious theory?

The only way to truly demonstrate a *Macbeth* curse would be to investigate if, on a clear and consistent basis, people really did experience *more* setbacks after quoting *Macbeth* than at other times. This would be difficult to detect, requiring years of careful, systematic study and observation. (And I'm pretty confident that it still wouldn't detect any actual curse.) In the meantime, remembering one single, mild setback on one occasion doesn't prove a thing.

Diamond in the Rough

How about my friend with the lost diamond, which reappeared after she prayed for its return? To her, this represented proof of divine intervention.

In this case, the diamond was very important to her and its return was indeed significant, so there was no *extra-large target*. Nor was there a *many people* trap, since she was just speaking about herself. Nor any kind of *false reporting*, since I'm confident my friend was speaking the truth.

Nevertheless, I took a somewhat different perspective than her. Why? There seemed to be a *many tries* trap here. Surely all of us—yes, including my friend—lose important items from time to time and wish strongly for their return. Sometimes, if we're *lucky*, we eventually find the item, sometimes in a very unlikely place. (Indeed, while I was revising this section, a diamond ring was found wrapped around a carrot in an Alberta farmer's field, 13 years after it was lost!)[1] At other times, items are lost forever. Given my friend's beliefs, I assume that she has often prayed for the return of lost items. In this one case, the item was subsequently recovered. But did that *always* happen?

I suspect that, when my friend prays for the return of an item, it is sometimes recovered and sometimes not. If that is so, but my friend only remembers the times when the item is found, then this is a form of *biased observation*.

On the other hand, if careful study showed that, on a clear and consistent basis, my friend was more likely to find lost items when she prayed for them than when she didn't—even after spending the same amount of effort searching for them—and if such searches were all systematically monitored and recorded to avoid any biased observation, then this would indeed constitute some evidence for the effectiveness of her prayers. But we cannot draw any conclusions from a single successful search.

By the way, this story also again raises the question of *cause*. Did my friend believe that the diamond was not in the fold of her shirt when it first fell, before she prayed? Was it on the floor and only moved to her shirt fold by divine intervention? Or had it fallen outside the car and then was magically retrieved and placed in her shirt fold? Or was it hidden in her shirt fold the entire time, in which case it's not clear what the prayers accomplished or changed. Like many stories about supernatural interventions, it is hard to pin down the details of what precisely is being claimed—which only increases the chance of misinterpretations and luck traps.

Karma Chameleon

And then there was that radio host, the one who found craps so interesting because of the associated "karma" she learned at a school in

Georgia, such as the rule that if dice fall on the floor, the shooter will probably lose on the next roll.

It is easy to imagine a "school" that collects hefty fees for doling out such gambling advice. The school, at least, is laughing all the way to the bank. But should we believe its claims?

My first thought was that dice falling on the floor is quite a rare experience. So it would take a long time to observe enough occurrences to draw accurate conclusions about how this does (or does not) affect the odds. Probably any one player—such as the radio host—would only see the dice fall on the floor a few times.

And this rarity in turn leads to questionable conclusions. Suppose you see the dice fall on the floor once. If the shooter then loses on the next roll, the karma theory is confirmed. If they don't lose, that can be ignored, since the theory isn't supposed to apply *every* time anyway. A classic case of *biased observation*. And without a long and difficult study of the patterns, there aren't enough opportunities to overcome the bias and see the bigger picture. It is much easier to simply believe the "school," remember any time that a shooter *did* lose after dropping the dice, and dismiss any time they won as insignificant.

Another way to think of this is as a form of *false reporting*—the school is telling you these karma "rules" as if they are facts. Does it have any clear evidence to support them? I sure doubt it. Nevertheless, its assertions can give people incorrect ideas. It seems that, with a convincing enough sales pitch, some people will believe almost any story about forceful luck that comes along.

There is also, once again, a question of cause. If the school is claiming that, by throwing the dice very carefully, it is possible to control how they will land, this seems very unlikely, but is at least physically plausible. But if they are talking about a "karma" force that decides whether you should lose, then that is a different matter. How could a pair of dice, unceremoniously falling onto the floor, possibly get together and agree to roll bad numbers next time to make sure the shooter loses? And, even assuming the dice had such magical powers, why would they do this? Why would they care if the shooter wins his next throw?

Indeed, all sorts of people make all sorts of claims about special "systems" meant to help you win at casinos. If they involve physically controlling the dice, then it is at least conceivable that they could help. But suggestions about what "karma" will lead to what results on which rolls at what times? There's just no way such things can possibly increase your odds.

Oh, and one final comment. If someone really did know how to consistently make money playing craps at casinos, then they could get rich from that. Why would they bother spending their time teaching for money at a craps "school"?

ESP in the Dumpster

And then there is that audience member whose dream allowed her to find her bracelet in the dumpster behind her apartment, which she took as proof of her special ESP powers.

Dreams are interesting and mysterious in many ways. For one thing, we don't always remember them. Indeed, we tend to remember only those dreams that have some significance. So even if we're trying to be careful, our understanding of our dreams is massively affected by selective observation.

I'm sure that my audience member has had loads of other dreams, equally vivid, that did *not* end up predicting anything real. Some of those dreams she didn't remember at all upon waking. And others she dismissed and did not remember for long—certainly not long enough to describe to a humble statistics presenter. Either way, a *biased observation*. Or, to put it differently, there must be many times when the audience member had dreams that suggested the location of some object. How many of them turned out to be accurate? Sure, this one dream was helpful in finding a lost item. But that is one dream, *out of how many?*

And what about my friend's question about whether ESP has to work every time just to exist. The answer is, it doesn't. But it does have to work more often than it would by luck alone. If the audience member's dreams really did allow her, significantly more often than random chance would expect, to learn information she could not otherwise have known, this could indeed constitute evidence of special powers—even if it did not happen every time. However, as with the other stories, to verify that would require lots of careful study. Just getting it right once, or even a few times now and then, doesn't prove anything.

So that is one possible explanation for the audience member's story. Out of all the times when she *could* have had a dream to tell her where a missing object had gone, she did have such a dream one time, just by luck alone. However, that might not be the whole story. A whole different reason for her success might also be possible.

Dream Me a Dream

When discussing dreams, there is always another factor at play too. Namely, the human brain. The human brain is incredibly complex, and not at all well understood. Even the best psychologists have only the vaguest notion of what part of the brain does what, and no one can come close to explaining how this strange, gelatinous blob is capable of such advanced high-level thinking, reasoning, remembering, emoting, and more. If I finally solve a math puzzle, I am pleased, but I have absolutely no idea what happened in my brain to find the solution. Furthermore, every mathematician has at times been stumped by a math problem, gone to sleep, and woken up knowing the answer. This suggests that, even while we sleep, our brains may be hard at work, solving puzzles and figuring out things that previously eluded us.

A woman once told me she had a dream that her close friend was upset because she had gotten divorced. Sure enough, a few months later, her friend did get divorced. Did the dream arise from special magic powers to see the future? Or had the dream caused the divorce? Neither. The woman had realized—subconsciously at least—that her friend's marriage wasn't going well. In her dreamlike way, she too had figured it out.

So, how did that audience member dream that her bracelet was in the dumpster? Simple: she figured it out! After all, she had searched everywhere in her apartment, without success. What had left her apartment? The garbage! And where had the garbage gone? The dumpster! When you think about it, the dumpster was the obvious next place to look. And she did think about it—in her dreams.

None of this makes the audience member's inference any less amazing. It is incredible how the human brain works, whether asleep or awake, and I'm sure many mysteries will remain for many, many years to come. But does this mean the audience member had magical ESP abilities, beyond the realm of science? Sorry, no, it does not. And that is why, as my talk title indicated, most statisticians don't believe in ESP.

Lottery Lunacy

And then there is my friend's girlfriend, the one who used her "intuition" about the spaces between the numbers on her lottery ticket to make her selection and was pleased that she matched four numbers out of six on her first try. Might there be luck traps at work there too?

For starters, could there be any *false reporting* involved? Well, we can't be completely sure her story is true. Perhaps she didn't really match four numbers, but only thought she did. Or perhaps, in her enthusiasm for her new method, she misremembered the result. Even more likely, perhaps she actually tried her new intuition several times, and only remembered her best result. All possibilities. But let's give her the benefit of the doubt and assume that she is remembering accurately, is reporting accurately, and did indeed match four numbers out of six on the one occasion when she tried her new method.

Is there a *many tries* luck trap? Perhaps. She had already explained that she "used to" use other methods involving the actual numbers, not the spaces between them. So presumably she had already tried a variety of other systems before hitting on her latest spacing theory. And if *any* of those previous attempts had produced a pretty good result, like four matches out of six, then she might well have told me about them instead. So there was indeed a *many tries* effect at play.

And what about the *many people* luck trap? In this case, there was only one person talking to me. However, over the course of my life, I have talked to lots of different people. Probably many of them have tried some sort of lottery "system" at some point or other. And if any of those systems had ever produced a promising result, someone probably would have mentioned it to me when we chatted. So there have indeed been other opportunities for other people to report similar claims, which creates a sort of *many people* effect too.

I computed later that the probability of matching four numbers out of six was about one chance in a thousand.[2] This isn't super low, especially considering how many lottery tickets my friend's girlfriend has probably purchased in her lifetime. So those matches were somewhat of an *extra-large target*. (By contrast, if she had won the lottery *jackpot*, her claim would have been much more dramatic.) Combined with the other luck traps, it seems that her achievement wasn't so unlikely after all. Did the girlfriend get lucky with her lottery pick? Yes, but only a little.

These were the thoughts that went through my head as I heard her tale. Simple, logical, and clear. But since she had asked me directly about her idea, and was standing right in front of me, the pressure was on. I awkwardly replied that I didn't really think such an approach could help increase her odds, and that her four matches were indeed probably just a coincidence. Then I quickly changed the subject, to avoid any further awkwardness.

But what did I *want* to say?

Well, part of me wanted to scream. The lottery numbers are chosen at random, from balls in an urn. There is absolutely no way for anyone to predict or influence the results by pondering a lottery card, whether they focus on the numbers, the "spacings," or anything else. How could my friend's girlfriend possibly have thought otherwise? And how could we have any kind of sensible discussion of the matter?

And yet, I have to acknowledge that many people feel the way she does, that it might perhaps be possible to use a system to figure out which lottery numbers to pick. And to such people, an early partial success, like four matches out of six, would only fuel their interest. The reality is that a majority of people, upon hearing our conversation, would probably side with her.

And why not? What could be more boring than saying that lottery draws cannot be influenced? And what is more exciting than rumours of a method that can beat the odds? Heck, I would be massacred.

I found myself musing even more about her claim that her system was based on intuition, so it "couldn't be tested." Now, that is not actually true. I imagined myself accompanying her on her next ten or twenty lottery purchases, having her pick the numbers using whatever intuition and spacings and patterns she wished, keeping careful track of how many winning numbers she matched, and comparing that winning record to what would have been expected according to the precise laws of probability. If she matched more numbers than chance would suggest, and if she could do so consistently, that would indeed offer strong evidence in support of her spacing system. At that point, I would be forced to take note of her apparent special powers. But if she didn't, if her results were pretty much the same as chance would predict, her system would be disproved once and for all.

However, a test like this would take quite a while and be rather awkward to implement. Worst of all, even if the experiment were carefully done, and showed no special abilities on her part, I fear that she—and others like her—might remain unconvinced. They might feel that my following her around and tracking her results had disrupted her intuition and ruined her abilities. Or that her abilities only appear at certain special times and can hardly be expected to work on the precise dates on which I chose to study her.

Most significantly, even if we conducted a test good enough that we all ultimately agreed that her lottery results were disappointing, she might simply decide to modify her system to a new one. Perhaps she would instead decide to add up all of the digits of the numbers, or measure their distance from the top-left corner of her lottery card, or some other scheme. In that case, we would have to start the study all over again.

It seems like such a simple situation: she believes in some weird system for predicting lottery results, and I'm confident that her system is invalid, and I even think it might be possible to test her system scientifically. And yet, it feels like it would be impossible for her and I to ever bridge our great divide.

Car Troubles

Finally, what about my student whose blind date was driving a car the exact same make, model, colour, and year as his own? What can we make of that?

It's certainly not every day that the person you're meeting drives exactly the same car as you. But how unlikely is it? Sometimes, a bit of thinking about the probabilities can be helpful.

For starters, there are about 250 car models on the roads in any given year,[3] but some of these are quite rare, so there are probably at most 100 different *common* car models. And, roughly speaking, each one is commonly found in about five colours. And, roughly speaking, they're probably about equally likely to be anywhere from brand new to, say, ten years old. So as a very approximate estimate, we might say that there is about one chance in $100 \times 5 \times 10 = 5,000$, or about a 0.02 percent chance, that any given pair of people will be driving the same exact car.

So, how can I explain such a striking coincidence? Easy! It's the *shotgun effect*, or *out of how many* principle, once again. Out of all the students I've discussed randomness with over the years, and all the different meetings and encounters they've had, and all the ways the encounters could have been surprising (same car, same shirt, same elementary school, etc.), I had heard about *one* in which a striking coincidence occurred. Not so surprising after all.

Ah, but I hear you complaining. How cold I am! How can I analyze this luck in such heartless terms? Don't I realize that such a magical coincidence must "mean" something? Surely it indicated that their date was "meant to be" and they would live happily ever after. Right?

Wrong. The student explained that, despite their identical cars, their date did not go well. They discovered that they had little in common, the conversation dragged on, they were both happy when lunch was finally over, and they never saw each other again.

Probability 1, Fate 0.

CHAPTER 8

Lucky News

O nce you start thinking about luck traps, you'll never read the news the same way again.

So many news stories, proclaimed with such drama and enthusiasm, are less meaningful than they first appear. Whenever you hear about something interesting, surprising, disturbing, or strange, your first thought should always be the same: Is it really significant, or can it be explained by luck traps?

More Cereal, Please!

With the clever heading "You are what your mother eats," a 2008 study in a prestigious biology journal made a startling claim: the more breakfast cereal a woman eats before getting pregnant, the more likely it is that her baby will be a boy.[1] Indeed, they divided 721 women in the early stage of pregnancy into three groups: 216 who had one or fewer bowls of cereal per week, 205 who had between two and six bowls per week, and 300 who had seven or more per week. They found that the first group had about 45 percent boys, the second group about 47 percent boys, and the third group about 56 percent boys. Hence their conclusion, which was picked up by news media around the world: the more cereal, the more boys.[2]

Could it be true? Did the results of their experiment really prove that eating cereal creates more male offspring? Or were the results just random luck?

The authors advanced various theories about how evolution had decided that male offspring are more valuable during times of plenty (represented by more breakfast), while female offspring are more valuable during times of scarcity. And about how skipping breakfast could depress a woman's glucose levels, which "may be interpreted by the body as indicative of poor environmental conditions" and so on. So the explanations for the phenomenon were all ready to go. The only question was: Was there actually a real phenomenon to explain, or was the variation just the result of luck?

What about luck traps? Well, one concern could be a *placebo effect*. Perhaps giving birth to a boy somehow "convinced" the mothers that they had eaten more cereal, when in fact they hadn't? (If that sounds far-fetched, consider that placebo effects have been shown to influence everything from pain reduction through acupuncture[3] to actual physical performance; one study found that runners took 2.41 seconds off their 200-metre sprint times after being told that their energy drink augmented running performance.[4]) However, in this case, the researchers had been careful to obtain all of the nutrition information *before* each mother knew the sex of her child. This seemed to rule out any psychological or placebo effects. What about a *lucky shot*? In keeping with the journal's standards, the authors had done a statistical analysis and claimed that there was less than one chance in a thousand that such a strong relationship between cereal and gender would have arisen by luck alone—not very likely.

So, if it's not a *placebo effect*, and it's not a *lucky shot*, then it must be a true effect, right?

Maybe not. Others quickly pointed out that there was a *shotgun* or *many tries* (or, more properly, *multiple testing*) effect here.[5] Namely, the authors hadn't only asked the mothers how much cereal they ate before getting pregnant; they had also asked them about their consumption of 131 other food items. And for each item, they considered how much was eaten the year before getting pregnant, during early pregnancy, and during later pregnancy. This means that, for each of the women, they had a total of $132 \times 3 = 396$ different intake measurements that *could* have shown an influence on baby gender. Out of all those possible influences, it is not surprising that *one* of them showed an influence in their study, just by luck alone. Seen from this broader perspective, their results had a simple, meaningless explanation, and their study didn't really prove anything about the tendencies of cereal to produce male children after all.

Beware the luck traps!

You Save My Life, I'll Save Yours

Some years ago, an amazing story made the news. Actually it was two stories rolled into one.

In the first part, a ten-year-old boy in western New York State accidentally got hit in the chest by a baseball bat—so hard that his heart stopped beating and he was about to die. Fortunately one of the baseball players' mothers was a nurse, and she was in the stands. She performed CPR on the boy. After a few tense moments, his heart started beating again, and *luckily* he recovered without difficulty.

The second part took place seven years later. The boy, now 17, was a volunteer firefighter, an Eagle Scout, and a kitchen worker at a local restaurant. A patron at the restaurant choked on some food and couldn't breathe. The boy was summoned. Using his firefighter training, he immediately performed the Heimlich manoeuvre on the patron. The food was dislodged, and *luckily* the patron could breathe again.

What makes these events so astounding? Remarkably, the choking restaurant patron was the very same nurse who had saved the boy's life seven years earlier. Wow![6]

There is an old saying: "The life you save may be your own." It was originally used as part of a campaign to prevent automobile accidents, imploring motorists to drive safely.[7] It was also the title of an unusual 1955 short story by Flannery O'Connor about a Southern woman marrying off her daughter, with unlucky consequences.[8] (It was adapted in 1957 into a television production starring Gene Kelly.)[9] But never was the expression more literally appropriate than in this case. If the nurse hadn't saved the ten-year-old boy's life on the baseball diamond, he wouldn't have been around to save *her* life seven years later. In a sense, one of the lives the nurse saved was indeed her own. What luck!

I was interviewed about this story for the television program William Shatner's *Weird or What?*.[10] On the one hand, I totally supported the main messages that this story would convey. You should indeed learn CPR and other life-saving techniques. And if people act kindly, they will indeed benefit in the long run. You *should* help other people, because (among other reasons) one day they might help you. And this story represents an amazing, fascinating, entertaining coincidence. On the other hand, I also wanted to try to compute the probabilities accurately and avoid any luck traps. So, what did I do?

First, I considered the following question: Suppose we have two specific individuals, A and B, both living in the United States. What is the probability that, sometime this very year, A will save B's life, and B will also save A's life? Using some simple approximations and estimates, I calculated this probability to be just one chance in tens of billions of billions—an unimaginably small number.[11] Viewed this way, the story seems extraordinarily remarkable, something that couldn't possibly arise by luck alone, and therefore must have some special meaning or supernatural cause.

But maybe not. I then considered the *many tries* luck trap. For one thing, there are many different years in which some such amazing coincidence might happen. And more importantly, there are lots and lots of different people in the United States. And there are even more *pairs* of people in the United States, even just counting those pairs who live in the same community and thus might perhaps happen to save each other's lives. (For example, if there are 15,000 people in a community who know CPR, then the number of different possible pairs of people who know CPR is already over 100 million.) Putting this all together, I calculated that the probability that *some* such event—of *some* pair of two people each saving the other's life—would take place at *some* point during my lifetime was actually approximately one chance in three.[12] Which means such an event isn't that unlikely at all. So despite my first impressions, it seems that this incredible story could indeed have arisen just by dumb luck alone.

But boy, what an amazing story of incredible luck it is.

Lightning Luck

Roy Sullivan was a park ranger at the long, narrow Shenandoah National Park in the state of Virginia. He is remembered for precisely one thing: he was struck by lightning a total of seven times. This fact is so astonishing that it is documented by Guinness World Records.[13]

Although Sullivan survived each of the strikes, they were not gentle.[14] In 1942 his right leg was burned and he lost the nail on his big toe. In 1972, and again in 1973, his hair caught fire. In 1977 he suffered chest and stomach burns. These were not minor, incidental lightning strikes, but rather serious encounters that caused great pain and injury.

What bad luck! What does it mean? Was it fate? Was he being punished? Was it part of some master plan? Maybe not. Once again, we can ask: Were any luck traps involved?

My first thought was that perhaps Virginia is a place with an especially high preponderance of lightning. But this does not seem to be the case. Indeed, the number of deaths per million people by lightning in Virginia during the period from 1990 to 2003 was 0.19, which ranks 27th out of the 50 states,[15] putting Virginia right in the middle. So *that* is not the explanation.

However, these lightning strikes still had some *hidden help*—Sullivan's work forced him to be outdoors, away from any buildings or protection, for long hours every day. He was thus much more likely to be struck by lightning than an average person.

But most of all, Sullivan was just one out of *many people* who could have been struck by lightning so many times. What we know is that, out of the hundreds of millions of people who lived in the United States during the 20th century, *one* of them got struck seven times. Most of them were never struck at all, and the remaining few were mostly struck one or two or three times, but not more.

When viewed within the larger context, with all of the *many people* involved, the story is not so surprising after all—though I'm sure that was of no comfort to poor Mr. Sullivan himself.

The Secret behind *The Secret*

In recent years, numerous celebrities, including Oprah Winfrey, Jim Carrey, Denzel Washington, Alan Arkin, Richard Gere, Jamie Foxx, Matt Damon, and Steve Harvey, have spoken about the power of the law of attraction—as outlined in Rhonda Byrne's 2006 bestseller *The Secret*—a special force that allows your thoughts to magically influence what happens to you. "You attract what you feel, what you are, what's on your mind," Washington explained. Harvey elaborated that "like attracts like. Whatever you are, that's what you draw to you. If you're negative, you're gonna draw negativity. If you're positive, you draw positive. If you're a kind person, more people are kind to you. If you see it in your mind, you can hold it in your hand."[16]

These celebrities cite the law of attraction not only as the basis for negative and positive emotions that come their way, but even for their career success. Carrey recalled that, as a poor struggling young actor, "I would visualize having directors interested in me, things coming to me that I wanted." This visualization, he feels, is what caused him to be cast in films. "You get it when you believe it," he explained.

Winfrey connected this "law" to her being cast in the movie *The Color Purple*. She explained that, upon first reading the book of the same name, she "was obsessed about this story" and desperately wanted to be in the film version. So, what happened next? "I auditioned. I don't hear anything for months. I wanted to be in this movie so much. I'm praying and crying. And as I'm [running] on the track... a woman comes out to me. And she says there's a phone call for me." That phone call, as you may have guessed, was about casting her in the film. Her conclusion? "I had drawn *The Color Purple* into my life."

So, is that the answer? Have these celebrities shown us the way to happiness and success? Do their triumphant tales, coming on the heels of their focused wishes, really prove that thinking controls reality? Many people think so—Byrne's book has sold over 20 million copies. But I don't. These testimonials come packaged with loads of luck traps, which change their interpretations significantly.

First of all, many of the claims have an *alternate cause*. It is indeed true that if you have a positive attitude, you will tend to speak and act more positively and helpfully, which will prompt other people to have a good feeling about you and respond more positively as well. Or, if you are kind towards others, then hopefully they will appreciate your efforts and treat you more kindly in return. I do agree with such connections and think they are a valuable life lesson (and one that I would do well to remember). But the outcomes can be explained by social interactions and conventions, combined with normal human reactions. They are not caused by any special force or energy emanating from the mind and directly controlling its surroundings.

Similarly, actors who imagine achieving great stardom are probably very committed to their craft. They spend lots of time studying and developing their skills. They pursue every possibility of advancement. They prepare hard for every audition and put their hearts into every single role. These actions all increase the probability of career success, but for very concrete reasons of hard work, being prepared, and seeking out opportunities. Again, a good life lesson, but an *alternate cause* that does not imply anything supernatural is at play.

And what of Winfrey's dramatic tale of being interrupted, while thinking intensely about the movie in the middle of a run, by a woman with news of her big acting break? Well, she already admitted that she was obsessed by the movie, thinking about it constantly. So of course, whenever the big news arrived, it would interrupt her in the middle of some activity she was taking part in while, yes, thinking about the movie. Clearly her story is an example of the *extra-large target* effect. (I don't mean to disparage Ms. Winfrey; while I was revising this chapter, she gave an excellent, passionate speech about gender and racial equality at the 2018 Golden Globe Awards.)[17]

Still, these are all just quibbles. The main story these celebrities are telling is that, after thinking hard about career success, they became big stars. How could this have happened, in so many cases, without the thinking magically causing the success? Easy. This is another *many people* luck trap. And who are the many people here? Why, all the young, aspiring actors around the world who are hoping to become famous, waiting for their big break, dreaming of becoming stars. There are surely many millions of such people—including some of my friends. They all think long and hard about success, daily hoping against hope that some big director will discover them or some fancy producer will lead them to stardom. Who wouldn't? Sure, some of them may have bad attitudes, giving in to thoughts of negativity and despair. But that still leaves millions who should have become stars according to this "law," but never did—instead, they eke out small, poorly paid parts for a while before eventually taking a day job and moving on. The only ones who ever appear in big documentaries discussing the reasons for their success or failure are the very, very few who do make it big, leading to an extremely *biased observation*. Seen in this light, the positive thinking alone had no magical effect, which is why it provided no help to most of the young, aspiring actors.

What I find most striking about these celebrity testimonials is their air of certainty. Like the rest of us, celebrities love magic and would rather believe in special, powerful forces that cause success than acknowledge that their success was the result of boring influences like hard work, preparation, talent, opportunities, and plain dumb luck. Despite the lack of scientific evidence, they all seem to feel that their New Age philosophy is established fact. "It's a law of physics," Arkin declared. "I don't know how anybody can disagree with that." Well, I do.

Nevertheless, belief in such laws is mostly just harmless fun—and perhaps even beneficial if it makes people feel more positive about their lives. But not always. One dramatic example occurred on Oprah Winfrey's television show. A woman named Kim Tinkham was so convinced by *The Secret* and its miraculous healing powers that she appeared on Winfrey's program in March 2007 to declare that, having been diagnosed with breast cancer, she had decided to "heal herself." She explained that three different doctors had urged immediate surgery, but that merely made her "mad," and she instead refused all conventional medical treatment (apparently with the encouragement of controversial naturopath Robert Young). Her decision put Winfrey in an uncomfortable position. On the one hand, she urged Tinkham to continue to "think as positively, think about attracting healing to yourself, think about the goodness that the healing will bring to yourself." But she also hedged her bets, adding, "Healing comes in lots of forms" and "I don't think that you should ignore all of the advantages of medical science," and cautioning that this alternative thinking "is not the answer to all questions." Tinkham stuck to her guns, explaining, "I'm making a choice," to which Winfrey replied, "That I respect." So, what happened? Tinkham died in December 2010, a tragic victim of her own false beliefs.[18]

Face Value?

A week after I submitted the first draft of this book, I came upon a recent computer-science paper that made a very dramatic claim. The authors had conducted an experiment in which computers used a *neural network* statistical algorithm to analyze more than 35,000 digital photographs of faces.[19] Their goal was to see if computers could learn to visually identify subjects' sexual orientation based on nothing more than a photograph. And what did they find? Well, their computers correctly distinguished between gay and heterosexual people with 81 percent accuracy for men, and 74 percent for women. Pretty impressive!

So, what does this mean? Had they demonstrated, once and for all, that sexual orientation can be detected based solely on appearance? If so, did that prove that sexual orientation is a biological property, controlled by genetics along with the shape of our faces? The researchers certainly thought so. "Those findings advance our understanding of the origins of sexual orientation," they insisted.

Not so fast. It was quickly pointed out that, among other flaws, the paper had used photographs from dating sites, so any perceived differences were as likely to be caused by gay and straight people choosing to present themselves differently (with respect to facial hair, makeup, glasses, etc.) as by any genuine differences in their appearance.[20] In short, even if the experimental results were valid, they came with a clear *alternate cause* quite distinct from any actual innate physical differences.

As one independent expert cautioned after reading the paper, "New discoveries need to be treated cautiously until the wider scientific community—and public—have had an opportunity to assess and digest their strengths and weaknesses."[21] Yes, indeed. Especially if, as in this case, they are marred by significant luck traps.

Seeing Red

Luck traps aren't just confusing for novices. They are debated in the highest reaches of academic research too.

Some psychologists recently published a study in the journal *Psychological Science*, making the dramatic claim that women were three times more likely to wear a red or pink shirt during their fertile days as during their non-fertile days, presumably due to an ancient instinctive drive to be more sexually attractive when sexual activity might lead to procreation.[22] The authors even included careful statistical analysis to back up their claims. "Female ovulation, long assumed to be hidden, is associated with a salient visual cue," they boldly declared, an assertion that was widely reported in the media.[23]

A statistician attacked their study on several grounds, complaining in an article in *Slate* about their small sample size (just 124 people, of whom 24 were students and 100 volunteered online), the definition of "fertile days" that was used, the inaccuracies of self-reported menstrual cycle timing, etc. But his main criticism was one of our luck traps, namely the *shotgun effect*. He said that the researchers could have tried different clothing items (besides shirts), different colours (besides red and pink), different definitions of fertile days (besides the "6 to 14 days after onset of menstruation" that they used), and so on. He said these choices provided excessive "researcher degrees of freedom" that the researchers could use to search and search until they finally came up with some significant-seeming finding.[24]

Now, the statistician certainly had a point. Suppose the psychologists had proceeded by asking their subjects lots of questions and then testing out all sorts of different hypotheses: Do women tend to wear green shoes on the day they start menstruating? Do they like orange hats in the middle of their cycle? Do they favour ponytails on even months during a full moon? By pursuing enough of these, the psychologists could create their own *shotgun effect*, in which *some* shot would hit the target by luck alone, signifying nothing.

So, did this luck trap make the research meaningless? Not necessarily. In the psychologists' response, they rebutted many of the criticisms. For example, they pointed out that they *only* asked their subjects about shirt colour, not other clothing items ("because we assumed that shirts would be the clothing item most likely to vary in color," they explained). Similarly they say they chose their 6-to-14-day time range *before* running any analysis, not afterwards. And they specifically set out to test the colours red and pink, which were already claimed by other researchers to be associated with sexual interest and attractiveness; they did not test every colour and then see which one mattered. Indeed, as a control, they checked and found that the effect of other colours was much less than that of red or pink and was not statistically significant, consistent with their theory that only the red/pink colour choice was meaningful. In short, the psychologists say that they tested only one specific hypothesis, not many. That is, they say they fired a single rifle shot, not a shotgun, and therefore they hit the "target" of statistical significance through legitimate means.[25]

So, what are we to make of all of this? Well, the claim that ovulation causes women to wear red does seem rather outlandish, and we should certainly approach it with skepticism. And yes, their sample size was fairly small and not truly random. So before drawing any definite conclusions, we should wait to see if their results are replicated by other researchers. But on the charge of *shotgun effect* and other luck traps, my verdict on these particular psychologists is: not guilty.

Supernatural Sisterhood

And speaking of menstruation (wow, I didn't expect to ever write that phrase), here's an age-old question: Are women who live together more likely to menstruate at the same time?

This "menstrual synchrony" effect is usually taken for granted. One recent article in *Cosmopolitan* begins, "If you've ever lived in a house full of women, you'll be aware of the hormonal chaos that can erupt once a month. Because as if one period wasn't bad enough, we girls tend to sync up."[26] The phenomenon is often attributed to such possible causes as pheromones excreted into the air by menstruating women, or the psychological effect of knowing that your friends are menstruating, or the influence of the moon's phases on menstrual timing. It has been hailed as a brilliant evolutionary innovation, forcing all females in a tribe to be fertile at the same time in the hope of preventing all of them from being impregnated by the same male. It has even taken on a feminist "sisterhood" symbolism, indicating the extent to which women naturally work together and empathize with each other. Just one question remains: Is it actually true?

Perhaps. A 1971 psychology study sure seemed to say so. It proceeded by arranging 135 students at a women's college dormitory into 15 groups of self-reported "close friends"—that is, the five to ten people they spent the most time with. Each student was asked her date of onset of menstruation each month. The study then computed how many days each student's onset was from the average of her close friends. And what did the study find? Well, the average distance from the mean menstrual onset date decreased over the course of the school year, from about six and a half days in October to about four and a half days by the time April rolled around. This decrease was found to be statistically significant, and seemed to confirm the theory that women who spend lots of time together start to menstruate in closer synchrony.[27]

End of story? Perhaps not. Some later papers criticized the methods of the original study, including the statistical analysis used, how different students' menstrual onsets were "matched up" to compute the differences, issues related to birth control pills (which artificially control menstrual cycles), and how irregular menstrual patterns were handled.[28] More importantly, other studies tried and failed to replicate the results, thus casting doubts on the original claims.[29] And a detailed study of menstrual patterns of the Dogon people in Mali concluded that different women's menstrual cycles had no clear effect on each other.[30] Indeed, it is now generally believed by scientists that the menstrual synchrony effect does not exist after all.[31]

On the other hand, despite this negative evidence, studies indicate that between 70 and 95 percent of women *believe* that menstrual synchronicity occurs.[32] Why would so many people believe in a phenomenon that most research indicates does not exist? Are some luck traps involved?

Yes, indeed. One issue is our *search for meaning*. It is much more satisfying to think that a fundamental biological phenomenon happens at productive times in solidarity with other people than to believe it is just random. This feeling is all the stronger if the associated meaning is not just biological, but also contains a social aspect of female solidarity.

Another issue is *biased observation*. Women are more likely to remember an occasion when they discovered that they'd menstruated at the same time as a friend or relative than to notice the times when they did not do so. This probably makes many people think that menstruation is synced more often than it really is.

But a further issue is an *extra-large target*. If we assume two women each menstruate once every 28 days, then the largest possible difference between their menstrual onsets is 14 days. (For example, if one woman's onset is 18 days after the other woman's, then it is only ten days before the other woman's *next* onset.) If they are completely random the onsets are equally likely to be anywhere between zero and 14 days apart, for an average of seven days apart. So having menstrual onsets within, say, four days is not so remarkable; it is only slightly less than average.

In addition, most menstrual periods last for approximately five days.[33] This means that whenever the menstrual onsets are within five days of each other, there will be some overlap of the corresponding menstrual periods. Even if the timings are completely random, this overlap will occur with a probability of about 5 out of 14, or 36 percent. In short, if two women are living together, then on average, their menstrual periods will overlap more than one-third of the time *just by chance alone*. That is a pretty large target, indeed.

Since menstrual overlap fulfills a need for meaning, occurs fairly often by chance alone, and is more likely to be remembered, it is not at all surprising that most women believe some sort of synchronizing force is involved—even though it probably isn't.

I'll leave the last word to Oxford anthropology professor Alexandra Alvergne, who noted that although "what we observe is nothing more than randomness," menstrual synchronicity is still "a popular belief. As humans we always like exciting stories. We want to explain what we observe by something that is meaningful. And the idea that what we observe is due to chance or randomness is just not as interesting."[34]

I couldn't have put it any better myself.

CHAPTER 9

Supremely Similar

S tories abound of striking similarities between different people, and they are usually told with claims of great meaning and significance. Whether the people are long-lost relatives, historical figures, or future spouses, there is something very satisfying about noting ways in which people are the same. However, with our greater understanding of luck, we now know enough to ask the all-important question: Are the claimed similarities really significant, or are they just the result of random luck, brought to our attention by one too many luck traps?

Like Father, like Daughter

I recently read a nice news story about a woman named Bernice Clarke, who lives in the territory of Nunavut in Canada's Far North. Clarke, who had never known anything about her father, searched for decades to get answers. Finally, after years of frustration, one connection paid off, and father and daughter were reunited. They had a joyful phone call in which they "were both giggling like teenagers." Clarke even declared that she had "won the DNA lottery." So far so good.

The story went on to marvel at the various similarities between the two, saying of father and daughter that "they have both worked for the airline industry, they're both entrepreneurs, they've both dabbled in acting, and love to travel."[1]

Ah yes, proof that our destinies are determined by our genes, and that a father and daughter share the same stories and goals, even if they've never met. Right?

Well, let's see. To evaluate the significance of this story, we need to ponder whether these similarities are actually meaningful, or are just due to luck traps.

The most striking thing, to me, about this list of similarities is what is *not* there. Did they live in the same place? No; she lived in the Far North, while he lived in Montreal. Did they speak the same language? Not really—he spoke English with a thick accent because his first language was French, which she did not speak. Did they currently work in the same sort of job, or have the same number of children, or like the same kind of music, or agree on their favourite food or colour? Presumably not, or the article would surely have said so.

So out of all the multitude of ways that they *could* have been similar, the article found four actual similarities. This is like the sharpshooter's *shotgun effect*—if you shoot enough pieces of lead, or consider enough personality traits, then eventually a few might hit the target by luck alone.

And what of these similarities? Well, surely *most* people would say they "love to travel"—who doesn't? And as for "dabbled in acting," who hasn't at some point had some role in a school play or something? Without further details, that doesn't sound significant either. Similarly, what does it mean to be an "entrepreneur"? Does it just mean that you sometimes buy and sell certain items to try to make a buck? Each of these characteristics is so vague that many, many people would share them by luck alone. This is like the sharpshooter's *extra-large target*—if you define your categories broadly enough, it's easy to hit the target.

This leaves us with one more characteristic, namely that the father and daughter have both worked in the airline industry. This isn't a triviality, since most people do not ever work in the airline industry. But it's not so surprising either. For one thing, the article doesn't specify what *kinds* of work they did, so presumably their jobs were different. Perhaps one was a flight attendant and the other a computer technician? Or one spent a summer as a customer service assistant, while the other had a part-time job helping with publicity? "Airline industry" is a pretty broad term, and a past job could be quite minor. Overall, this is again quite a large target to hit.

So that's it. Out of all the investigating of this father-daughter pair, the only true similarities involved *extra-large targets* and a *shotgun effect*. Not so amazing after all. Don't get me wrong—I'm delighted that this woman finally found her father, which led to comfort and joy. But don't push your luck by claiming that the pair are more similar than they really are.

You Guys Could Be Brothers

While writing this book, I came upon an interesting newspaper article about a Calgary train operator named Avrum Gordon, who was adopted at birth and finally tracked down his birth father at the age of 48. This led to his visiting some other newly found biological relatives, who received him warmly with a sign saying, "Welcome to the family, Avrum."[2] So far, so good—a heartwarming story about a successful late-in-life family reunion. But not particularly surprising, since millions of adopted children have successfully contacted their birth parents.[3]

Where the story gets interesting (at least from a luck standpoint) is what happened next. During his initial hour-long phone chat with his biological father, Avrum learned that he had a younger brother named Chris Dyson. It seems that, after giving Avrum up for adoption in their mid-teens, his parents kept dating, eventually married, and had two more sons, whom they kept. And amazingly it turned out that Avrum already knew Chris—in fact, they were "pals at work" as well as friends on Facebook. "Well, that's one hell of a coincidence," the father accurately summarized.

If that was the end of the story, then I could appreciate it at face value. After all, who doesn't love a story of surprising connections, like a long-lost brother turning out to be a work pal?

However, the newspaper couldn't leave it at that. It had to embellish the story with additional meaning and connections and coincidences, most of which do not stand up to scrutiny.

For example, the reporter wrote that Avrum had been "feeling isolated" before meeting his newfound family, and that "there was always something missing," thus lending extra emotional significance to the reunion. Well, who doesn't sometimes feel isolated? This is such an *extra-large target* that no more need be said.

Chris said that even when he first met Avrum, long before he knew their family connection, "There was something I really liked about him and I felt this kinship." That sounds sweet, but the article also admits that they were "not really close friends," so at best this claim is rather exaggerated. Chris also says that when Avrum's wife first saw Chris's Facebook photo, she noticed a striking resemblance and remarked, "You guys could be brothers." However, the newspaper article admits that this story "varies depending on who tells it." So this all sounds mostly like *false reporting*.

The newspaper also comments, "The family has since learned of a pile of coincidences." Mostly this consists of two facts. First, as youths, the two sons briefly lived about a block apart in Winnipeg. This is indeed something of a surprise, so I'll give them that one. (Though even there, the article has to overhype the connection, adding, "It is entirely possible that they played in the same park at the same time, maybe even with each other," despite no actual evidence of this.) And second, Avrum now lives in a neighbourhood of Calgary, Cranston, that is "the next community over" from the neighbourhood where their dad once lived. This seems like an *extra-large target*, given the larger distance (about four kilometres) between them and the rather tenuous nature of the connection (it wasn't the two brothers themselves, but rather one brother and a father—and they weren't even there at the same time). One brother tried to build up this coincidence by telling the other, "If the Sobeys would have been open before they moved, you would have shopped at the same grocery store," but this seems like a pretty weak attempt.

The article further adds that "all three brothers love to cook," and that "all three brothers report feeling happier since the reunion"—surely more *extra-large targets*. It also notes that Chris and his father are both science-fiction fans, which sounds impressive until you realize that this means Avrum presumably isn't. (And speaking of differences between the brothers, the article admits that Avrum's initial job was as a dental technician, while Chris worked in film, a very different career choice. But you can bet that if those choices had been at all similar, they would have been hyped too.)

Oh, and what about the fact that they were "work pals"? Well, it turns out that Avrum had already left his train job before learning of this connection, so they were at best *former* work pals. And Chris already admitted that they were "not really close friends," not truly "pals" at all. So, not terrible, but a little bit of *false reporting* there. And as for being Facebook friends? Well, I have over 500 Facebook friends, some of whom I have never even met, so I can say with confidence that this is another *extra-large target*.

In short, it's a nice story of a happy family reunion, with a few genuine surprises (the brothers already knew each other and once lived about a block apart), but with an awful lot of extra hype when the reporter just couldn't leave a good story alone.

Presidential Similarities

Over the years, the legend has grown about eerie similarities between US presidents Abraham Lincoln and John F. Kennedy:[4]

- Lincoln was first elected to the US Congress in 1846, and Kennedy was first elected to Congress in 1946, exactly 100 years later.
- Lincoln was first elected president in 1860, and Kennedy was first elected president in 1960, exactly 100 years apart.
- Lincoln was inaugurated in 1861, and Kennedy was inaugurated in 1961, also exactly 100 years apart.
- The names Lincoln and Kennedy each contain seven letters.
- Both presidents were particularly concerned with civil rights.
- Each of them lost a child while living in the White House.
- Both presidents were shot in the head.
- Both presidents were shot on a Friday.
- Both presidents were smiling immediately before the assassination began.
- Both presidents were assassinated by Southerners.
- Lincoln was shot by John Wilkes Booth at Ford's Theatre; Kennedy was (allegedly) shot by Lee Harvey Oswald in a Lincoln automobile, made by Ford.
- Lincoln had a secretary named Kennedy, who warned him not to go to Ford's Theatre (where he was shot). And Kennedy had a secretary named Lincoln, who warned him not to go to Dallas (where he was shot).
- Both presidents were succeeded by their vice-presidents, who were both named Johnson.
- The successors' names, Andrew Johnson and Lyndon Johnson, each have 13 letters.

- Their (alleged) assassins, John Wilkes Booth and Lee Harvey Oswald, were each known by three names comprising a total of 15 letters.
- Booth was born in 1839, and Oswald was born in 1939, exactly 100 years apart.
- Booth ran from the theatre and was caught in a warehouse. Oswald ran from a warehouse and was caught in a theatre.
- Both Booth and Oswald were assassinated before their trials.
- Presidential security was heavily criticized, after each assassination, for being too lax.
- There were theories that both Booth and Oswald were part of greater conspiracies.
- After Lincoln's assassination, the nation experienced an emotional convulsion. After Kennedy's assassination, the nation experienced an emotional convulsion.

This sure is a long and impressive list of coincidences. But what does it mean? Are the similarities just the result of dumb luck? Or are they meaningful results from magical, supernatural forces? Are there luck traps involved? How should we evaluate the various elements of luck involved in all of these coincidences?

Well, first of all, some *false reporting* is involved. Booth was actually born in 1838, not 1839. And Lincoln did not have a secretary named Kennedy. Don't believe everything you read! There is also no evidence that Kennedy's secretary warned him not to go to Dallas. Oh, and those assassins? Apparently they were not normally known by all three names until after their deaths. Furthermore, Booth was caught in a barn (in fact, a tobacco shed), which is not traditionally referred to as a "warehouse." And he was from Maryland—not normally considered the South—to boot.

There are also clearly some *extra-large targets*. What president ever was not "concerned with civil rights"? What president would not smile at a public event? What is the most likely spot to be fatally shot, if not the head? What president did not, in a difficult time, receive some sort of "warning" not to go somewhere (in fact, both of these presidents were *frequently* cautioned not to travel)? What assassination doesn't lead to some conspiracy theories? What security wouldn't be criticized after allowing a president to be assassinated? And what presidential assassination does not lead a nation to some measure of "emotional convulsion"?

The *extra-large targets* don't end there. Loads of surnames consist of seven letters,[5] and Johnson is quite a common name, so those coincidences aren't so surprising. Also, presidential elections are only held once every four years, so the time between them couldn't be 99 or 101 or 102 years, which makes 100 years somewhat less surprising too.

Furthermore, some of the coincidences are *facts that go together* to some extent. Most obviously, presidents are always inaugurated the year after their election, so if two of them were first elected 100 years apart, then of course they were also inaugurated 100 years apart. Similarly, it's reasonable that a president might first be elected to Congress approximately 14 years before being elected president, so if one such election is 100 years apart, then it's not so unlikely that the other one would be too. So these bits of extra evidence aren't that much extra after all.

Also, some of the "similarities" are not as similar as they first appear. For example, although both presidents lost a child in the White House, one was a stillborn baby (Kennedy's), while the other was an 11-year-old son who died of typhoid (Lincoln's). And Oswald was killed at close range by a private citizen (Jack Ruby) two days after his arrest, while Booth was shot by a federal trooper (Boston Corbett) 11 days after the incident, after being surrounded and failing to surrender. Pretty different, really, corresponding to either *false reporting* or an *extra-large target*, depending upon your perspective.

What about days of the week? It is true that both assassinations took place on Friday, and this has just one chance in $7 \times 7 = 49$, which seems somewhat unlikely. However, this is an *extra-large target*. If the assassinations had both happened on Thursday, that would have been equally surprising. Or Wednesday. Or any other day. The chance that they would both be on *some* same day of the week is one in 7, not one in 49. So it is unlikely, but not overly unlikely.

And is there some *shotgun effect* at work here too? Yes, indeed. For example, the surnames Lincoln and Kennedy each contain the same number of letters, but their full names do not. And, the full names John Wilkes Booth and Lee Harvey Oswald both contain the same number of letters, but their individual surnames do not. Similarly, the two presidents were not born or killed 100 years apart, nor in the same month, nor on the same day of the month, nor in the same state. Nor were the assassins born in the same month, nor did they die in the same month or state. And so on. There are many other possible coincidences that *could* have occurred, and would have been added to list if they had, but they didn't.

With all these provisos, is the list of actual similarities between the two presidents interesting or cool at all? I say: sure it is! The 100-year differences in their initial elections is still somewhat interesting, their successor vice-presidents sharing a surname is somewhat unexpected, and the Ford's Theatre/Ford car coincidence is rather striking. Overall, the coincidences can partially be explained away, so they are not *so* surprising, but they are still fun to chat about.

But do these coincidences actually *mean* something? No, I don't think so. No big pronouncements can be made about them. They were not part of some grand design, with a hidden meaning. Rather, well, they were just dumb luck.

The Case of the Haunted House

Next, to change things up a little, we hear about the latest case from that famous, crusty, lovable, mathematical, fictional probability detective, Ace Spade, PPI (first introduced in chapter 9 of my previous book, *Struck by Lightning: The Curious World of Probabilities*). Warning: serious, sober-minded readers may wish to skip this chapter.

The years since I'd saved Baker's bacon had been good ones. Baker's casino had been losing tens of thousands of dollars, and he was more puzzled than a high school slacker at a geometry midterm. After investigating, I'd fingered a dirty roulette player in cahoots with Baker's own fiancée, saving the casino and putting Baker on the path to prosperity. Baker was so grateful that he spread the word to other casino owners like a supercritical branching process. Before long, my business was humming along like an iterative equation. Card palmers, roulette scammers, poker cheaters—I routed 'em all out and collected my hefty fees. Doris and I had settled down in a nice house by the turnpike, and our delightful daughter, Denise, had just turned three. Life was sailing along like an unbounded diffusion.

One day, the phone rang, and Doris picked up. "Ace Spade, Probabilistic Private Investigator," she sang with her usual chirp that used to annoy me but now only increased my love. "How may I help you?" There was then a pause, after which she said, "Uh-huh." Another pause, and then, "Oh, really?" A few more seconds, followed by "What do you mean, 'haunted'?" And finally, "Okay, I'll ask him."

With that, Doris put her hand over the phone and whispered to me, "We've got a live one here. Something about a haunted house. The guy seems pretty agitated."

I started to protest. It's one thing to deal with casino owners, who might be hotheaded but at least understand dollars and sense. But superstitious flakes afraid of haunted houses—that's another story. Before I could object, Doris added, "We did have that cancellation tomorrow, so your day is open. It couldn't hurt to at least hear him out." She was being sensible as usual, so I found myself agreeing to meet him at my office the next morning.

He arrived right on time. Nervous and uncertain, with a jacket baggier than the convex hull of a fractal, he introduced himself as McGee. "I own a large house on Elm Street," he explained with a whine. "I rent out some of the rooms for a little extra cash. Seemed like a simple enough plan. But nasty things have been happening there, Mr. Spade. Real nasty!"

Now, I know a few things about luck. Sometimes it's bad, and sometimes it's good. Usually, despite what some might wish, it doesn't "mean" anything. So, a few unfortunate accidents at some house somewhere? Big deal. Could happen any time. Doesn't prove anything. Nothing to see here. "Listen, McGee," I began, "I appreciate your stopping by, but there's really nothing I can do to—"

I was interrupted by McGee waving a wad of green in my direction. "I can pay you a $2,000 cash retainer in advance," he explained.

"Please, sit down," I immediately offered with a big smile. "Please, tell me all about it."

McGee handed over the dough, which I quickly pocketed. Then he started rambling on about cracked bottles and sprained ankles and broken wrists and irate tenants and angry arguments, to the point where I couldn't keep track. "Look," I suggested, "we're not going to sort this all out here. Why don't I come by the house on . . ."—and here I checked my calendar—"Thursday, when I'll check everything out. Have all your tenants there too."

Thursday came fast, and I found myself ringing the bell of a large but ordinary-seeming suburban house on Elm. McGee himself answered the door and quickly ushered me inside. I was impressed by the brightly painted walls, grand chandelier, fancy tiled floors, flowing curtains, and other nice touches. "Had the whole house renovated last year," McGee said with pride. "I wanted it to look nice for my tenants. I run a respectable home here, Mr. Spade. But lately everything has been so . . . so terrible. The place is cursed, I tell you. Cursed!"

I took a look around. The ground floor contained a shared kitchen and living room, as well as McGee's own bedroom with ensuite bathroom. The second floor had two bedrooms, another bathroom, and a small laundry room. This floor was equally nicely renovated, with bright yellow walls, elaborate trim, glistening rainbow-coloured floors, shiny water taps, and a large, deep bathtub. The third floor was smaller and more rustic, containing two small bedrooms, each with sloped ceilings and plain wooden walls. The unfinished basement was poorly lit, with creaky stairs and cobwebs, a moistness that had gradually softened the many scattered cardboard boxes, and walls with exposed drywall and nails sticking out here and there. Overcome with an uncomfortable feeling, I climbed out of the basement and back to the living room, where McGee had assembled his tenants.

"All right," I began. "I need facts. You say some bad things have been happening here?" The tenants all nodded fervently, practically in unison. "Okay," I continued, "let's hear it. McGee, I'll start with you. Those complaints you were telling me about—did those things all happen to you?"

"Oh, well, uh, no," he replied. "Not to me personally. The spirits don't haunt me directly. What they do, though, is attack my tenants, so they're all ready to move out now, after which I won't get any rent, so I'll end up broke. The spirits pick my tenants one by one. They break objects, cause falls, rain destruction upon us. They—"

"Hold it down," I interrupted sharply. "You're just repeating your impressions of other people's stories. That can lead to *false reporting*." McGee glared at me, so I added, "No offence," which didn't mollify him at all.

I continued, more gently, "All I'm saying is, let's hear the stories from the people they actually happened to."

Everyone agreed, and a tall, thin fellow spoke up first. "My name is Carlson," he began in a snooty voice with a hint of an English accent. "I live on the second floor, in the *large* bedroom there." He seemed awfully proud of his bedroom size. "I'm a wine connoisseur," he went on. "I can distinguish excellent from mediocre wine with a mere sniff, and after three sips I can tell you precisely which wine will be next year's top seller. I've had my reviews published in all the major newspapers, and am starting a wine investment business that will soon make me unimaginably rich."

If you were rich, I thought, then you wouldn't be living in a rooming house. He must have seen my reaction, because he quickly added, "You may laugh, but I'll be the one laughing soon, I assure you."

I tried to move the conversation along by asking him why he thought the house was haunted. "Ah yes," his voice rose, "I will tell you. Last month, I tasted the most exquisite Aglianico. I won't tell you which one," he gave a conspiratorial wink, "but it is made by a tiny vineyard. Totally unknown. Doesn't even appear in the major listings. And yet, I am sure that by next year, wealthy drinkers from Naples to Napa will be beating down my door to buy it. That is, after they discover that I have become the majority owner. Ha ha ha!" With that, he laughed so loudly and triumphantly that he threatened to shatter the beautifully appointed living-room windows.

"And the haunted part?" I asked him with more than a touch of impatience.

"Ah yes," he returned to Earth. "Well, the very night I made my discovery, and tasted such perfection, I went to wash out my favourite wine glass. As usual, I used my special foam brush—no soap, of course—and scrubbed it the same way I always do. Then suddenly it shattered, right between my fingers. Glass flew everywhere. I cut my hand! Why would my glass choose to break at the very moment of my triumph, if not for evil spirits trying to stop me?"

I started to thank him for his explanation, but before I could, he continued, "And that's not all. As I tried to deal with the shattered glass, I knocked over the unfinished bottle of wine, which cracked in two and spilled all over the kitchen floor. It was a total mess. The purple stains still haven't come out! Those spirits are trying to foil my plans, but I won't let them!" He was getting very agitated now, almost hysterical, and it was all I could do to get him to sit down and stop talking. "And when I went to deal with the mess, I noticed a nasty cut on my hand," he finished in a sad voice before trailing off.

Next up was Teena, a short, pretty brunette who lived in the smaller second-floor bedroom. Her right arm was in a large cast. "I was working late on my thesis—I'm doing a doctorate in English lit—and I got hungry. I didn't want to stop writing, and I didn't want a pizza—too unhealthy—so I ordered a gourmet salad to be delivered from the vegan place up the street. Thirty minutes later, I was lost in thought at my computer when I heard the doorbell ring. And ring, and ring. Followed by a loud, aggressive knocking. Then, just as I started down the stairs, I flipped over and fell forward headfirst. In desperation, I put out my hands to break the fall, and broke my wrist! Now I'll be in a cast for six weeks, and I can only type with one hand. It's the evil spirits trying to delay my thesis, but I won't let them!"

I thanked her and turned to the next tenant, but Teena wasn't done yet. "When I hobbled to the door, in obvious pain, the delivery guy didn't even ask what was wrong. He just gruffly handed me the salad, made me sign the credit card slip—fortunately I'm a lefty—and barged off." I nodded, and she went on. "I was shocked by his rudeness, and tried to say something. But he kept walking and actually started to laugh at me. A cold, cruel, evil, haunted laugh. He was possessed, I tell you. Possessed!"

"Okay," I began, but she spoke again. "And in the morning, when I'm trying to sleep, I sometimes hear a weird, slow, creaking sound. Creak, creak, creak! There's villainy in the air!"

I tried to give her a look that was sympathetic, but would also end her testimony. But she snuck in one last remark. "Have you ever tried to type with one hand, Mr. Spade? It's brutal! The spirits are mocking me. I'll never finish my thesis like this. You've got to do something!"

Finally I managed to get to the next tenant. His name was Kondar. A large, strong college student, he clearly spent more time on the gridiron than in the classroom. "I ain't afraid of hard work, Mr. Spade," he began. "Last year, I played football, baseball, and b-ball, plus I lettered in track. I never missed no practice in none of 'em. You can ask my coaches!"

I had less interest in his coaches than a non-additive measure space, so he went on. "Then, last month, I'm ready for football practice, so I take off from my room as usual. But halfway to the front door, something gives way. My foot slides around hard and I feel this tugging, and it's sore as hell." With this, he stood up to show me his left ankle, and I noticed for the first time that he was limping slightly. "Doc tells me the ankle is sprained, and I can't run on it for three more weeks. I'm gonna miss our home opener! What's fair about that?"

Once again, there was more coming. "Then, when I finally get home from having my ankle looked at, I get a glass of milk for comfort, and it tastes sour. It's gone completely off. Disgusting, man! Why are the spirits doing this to me?"

Barely pausing, he went on. "Oh, and on New Year's Eve, I got this big, ugly cough that just wouldn't quit. I could barely breathe for, like, three weeks!"

He started to sit back down, but then remembered something else. "Oh, that ain't all. Over the holidays, the school gym was closed, so I had to do my dribbling reps in the basement. And one day, while I was swingin' my gym bag into the corner, my leg pushed against a big metal nail that came outta nowhere. Tore my trousers, made my calf bleed, hurt like hell." I thought he was done, but he added one further fact. "Good thing it was holidays, 'cuz I could hardly run to save my life."

Kondar gingerly sat back down, and the fourth tenant finally spoke up. She was slight and elderly, with bright grey hair in a tight bun. She introduced herself as Mrs. Stewart. "I'm no' norrrmally one ta compleen, Mr. Spaaade," she began in a gentle Scottish burr. "I've lived on the theerd flooor here for yeeeears, without any troooble." She smiled sweetly in my direction, but I saw fear in her eyes. "I'm no' as yooong as I uuuused to be, so I'm a bit slooow goin' up the steeeeers, but I do no' meeeend the delay. Tha' is, until a few weeeks agooo, anyway."

She paused, and I had to motion for her to continue. "One Toooosday evenin', I was reteeerrrning hooome from a brrridge game, all tiiired out after a looong daaay. I walked slooowly and quietly up to my rrrooom. I was using my cane as uuusual," she explained, holding up a large brown cane with a wide four-point base. "All o' the suudden, the cane pulls away from me and goes flying off, leaving me heelpless! I nearly toombled down the steeers, I did. I coulda brooken me neck. I coulda died!"

She had started out calmly, but she was getting more agitated as she spoke. "I'm no' a young woman, Mr. Spade. I canna take a chance on just fallin' this way and tha'. I won't take the chance o' livin' in a haunted house any longer. I've got to move out, Mr. Spade! Let the spirits pick on someone else, I say!"

At this, the others joined in with agreement. "It's true," one shouted. "No, it isn't fair," another added. "We didn't do anything wrong," one complained, "so why should we have to put up with this?" "This place is haunted," another repeated for emphasis. The others repeated the refrain. "It's haunted." "Haunted!" "Spirits are about!"

It was a lot to keep track of, so I excused myself and walked out to the front porch. There, I took out my trusty black notebook and jotted down what I remembered. There was a lot of hype in that house—angry words about spirits and haunting and evil—so I tried to separate out the facts. What actual bad things had occurred?

Here's what I wrote:

Carlson
Broken glass
Cracked wine bottle
Purple stains on floor
Cut hand

Teena
Aggressive knocking
Fell down stairs
Broken wrist
Rude delivery guy
Heard evil laugh
Delivery guy seemed possessed
Strange creaking sounds

Kondar
Foot slid
Sprained ankle
Sour milk
Torn trousers
Bleeding calf
Persistent cough

Mrs. Stewart
Cane flew away

    I carefully read the list several times, and started to think. What did all these negative experiences mean? Was the place really haunted? Or just going through some dumb, rotten luck? Or were other factors at play? Inside the house, McGee and the tenants were all in a lather, stoking each other's fears like a positive-reinforcement mechanical spring. I had to figure something out, and fast.

    My first thought was that some items on the list were *facts that go together* and could be combined into one. Sure, Carlson cut his hand, but that was only because the wine glass broke in his hand. And sure, he saw purple stains on the floor, but only because he'd cracked and spilled the wine bottle. And on it went. The aggressive knocking that Teena heard, and the dismissive laughter that upset her so, and her feeling that he was possessed, were all directly caused by the rudeness of the delivery guy. And her broken wrist was a direct result of falling down the stairs. In the same way, Kondar's sprained ankle was the result of his sliding foot, and his torn trousers and bleeding calf were both from getting stabbed by the same nail.

    Once I'd combined the items that went together, I was left with this:

Carlson
Broken glass (which cut his hand)
Cracked wine bottle (which stained the floor)

Teena
Rude delivery guy (who knocked aggressively and laughed cruelly)
Fell down stairs (which broke her wrist)
Strange creaking sounds

Kondar
Foot slid (which sprained his ankle)
Sour milk
Stabbed by nail (which tore his trousers and made his calf bleed)
Persistent cough

Mrs. Stewart
Cane flew away

It was still too much to dismiss, but I wasn't finished yet. Carlson had cracked the wine bottle right after breaking the glass and cutting his hand. And both accidents came when he was all excited about his Aglianico get-rich-quick scheme. It wasn't so surprising that he would break things in such a state. I decided to pay no further heed to Carlson's clumsy wine evening.

Similarly, Kondar's nail encounter was no surprise in such a dark and disorderly basement. And his cough was probably the direct result of the exposed walls and foul, moist air that he breathed hard while dribbling his basketball down there. In both cases, the condition of the basement was a perfectly reasonable *alternate cause*. So these troubles provided a clear warning about the basement, but didn't relate to the rest of the problems. I decided to exclude them from the remainder of my analysis.

Oh, and Teena's strange creaking sounds? No surprise there. Most old houses—even after extensive renovations—have some creaky floorboards. Plus, her room was right below Mrs. Stewart's—an elderly woman who mentioned getting tired in the evening. If Mrs. Stewart tended to wake up early and walk around a little—well, that explained the creaking sounds right there. So forget them.

A few other items could be similarly ignored. Who hasn't had a rude delivery guy now and then? Especially a vegan! That was simple human nature, evidence of nothing. And as for Kondar's sour milk, well, he was an eager, forgetful college-student/jock—I would wager a gold-plated quincunx that his milk went sour more often than attempts to prove the Riemann hypothesis. I quickly struck them off the list too.

Finally I was left with this:

Teena
Fell down stairs (which broke her wrist)

Kondar
Foot slid (which sprained his ankle)

Mrs. Stewart
Cane flew away

The pattern was becoming clearer. Once the irrelevant items were removed, all that was left were three physical accidents. And each of them involved some sort of sliding or falling. Coincidence? Maybe. But maybe there was an explanation waiting to be found.

I tried to think further. Where had these events taken place? Teena lived on the second floor and said she fell just as she started down the stairs. So she must have been at the top of the staircase between the first and second floors.

And what about Kondar? He lives on the third floor. He said he was halfway to the front door when his foot slipped. And where is halfway between a third-floor bedroom and a front door? Well, somewhere on the second floor. Interesting.

Then there was Mrs. Stewart. She was returning home and walking into her room when her accident occurred. No, not into . . . "*up to*" her room, she had said. So she must have climbed some of the stairs before her cane slipped. But how many? I decided to check.

I pushed my way loudly back in through the front door. The tenants had been arguing with each other and with McGee, getting more and more upset about their common misfortune. They turned when they saw me, but they didn't seem optimistic that I could affect their otherworldly demons.

I got straight to the point. "Mrs. Stewart!" I called, loudly enough that she jumped. "When your cane slid, which floor were you on?"

Mrs. Stewart recovered enough to consider this. "Well," she began, "I'd taken off me jacket and starrrted slooowly climbing up the steeeers . . ." She considered for a few seconds before adding, "I suppooose I'd made it up to the second flooor, but only beeeerly, since—"

"Aha!" I exclaimed, without even waiting for Mrs. Stewart to finish. "So all three accidents happened right around the top of the first staircase!"

My triumphant declaration didn't have the desired effect. Rather, everybody started objecting.

"These were no accidents, these were evil spirits!"

"*Three* accidents? Why, there were a lot more than three!"

"The whole house is haunted, not just one spot!"

I ignored them all and continued my explanation. "And what is at the top of the first staircase? Why, a rainbow-coloured floor. Different colours made with different materials and different paints and different finishes. And one of them," I finished with conviction, "is more slippery than a proof by contradiction."

They were briefly puzzled by my analogy, but quickly returned to the matter at hand. "Slippery? What do you mean?" McGee demanded. "This is what I'm paying you for? Why, I paid top dollar for my renovations last year. Nothing slippery about it!"

"Let's see," I replied, quickly bounding up the nearby stairs. At the top, rather than walk on the dramatic floor, I bent over and started feeling it with my fingers. A bright yellow region felt solid enough. A dull red area was similarly supportive. But a small deep-orange strip running on an angle felt strangely smooth, like nothing could ever stick to it—not foot, not shoe, not cane.

"That's it!" I declared, showing the orange strip to McGee. He was skeptical, but felt it too, and admitted it was a lot more slippery than he would have guessed.

"What happened," I explained, "is that every once in a while, a tenant hurrying by this spot—say, collecting a salad, or going to football practice, or returning from a nice game of bridge—would happen to step entirely on this orange bit. Just bad luck, you know," I continued with a wink. "And once they did, the support would give out on them, causing a slip or sprain or fall. And *that* is the cause of your problems."

"Not mine," Carlson quickly objected. "I was nowhere near that orange strip when I broke my glass and cracked my bottle and cut my hand."

"No, not yours," I admitted. "Some of your misfortunes just happened through bad luck, like an excited wine investor scrubbing a little too hard, or a vegan delivery guy who woke up on the wrong side of the bed. And a few others were caused by that death trap of a basement—which should be fixed up right away," I added, glaring at McGee. "But the truly surprising ones, the ones beyond what might happen by chance alone—they were engulfed in a slippery tale of orange."

Carlson wasn't impressed, but by this time McGee had come around. "You're a genius, Spade," he said, shaking my hand. "I'll get the boys to add a rough finish to make the orange less slippery, and then we can all return to our earlier happy ways. I knew a probabilistic private investigator was the key to our happiness!"

A moment later, still engulfed in joy, McGee turned to me and admitted, "Oh, and I guess I'll patch up the basement too."

That evening, I returned home after a long but satisfying day, just as Doris was putting beautiful little Denise to bed. "How did it go?" she asked.

"Just fine," I replied with a smile. "I think I can safely say that, using the logic of luck, I have managed to solve, once and for all, the case of the haunted house."

At that point, Denise raised her little head from her pillow and turned to look at me with a funny expression on her face. "Don't be silly, Daddy," she giggled, wriggling her tiny nose. "Everybody knows that there's no such thing as a haunted house!"

CHAPTER 11

Protected by Luck

Our world is full of dangers. At any moment, a criminal could rob us. A terrorist could drop a bomb. The airplane we're flying in could crash. Our child could be abducted. So many bad things could happen, to such terrible effect. What can we do about it?

Well, to some extent, we can take precautions. We can lock our doors, avoid dark alleys, stay out of war zones, and keep an eye on our children. But such steps only protect us so much. Fortunately there's another way to avoid trouble: *luck*.

Honesty Abounds

In the movie *European Vacation*, Chevy Chase hands his video camera to a random passerby and asks him to take some footage of his family by a Paris fountain. The man suggests that they remove their shoes and step right into the fountain itself, which they eagerly do. Unfortunately it is all a trick, and the man runs off with their camera. (And, this being a Chevy Chase movie, footage of his wife found in the stolen camera ends up on a billboard in Rome . . . but never mind that.) Wow, we think—what bad luck.

However, I think this scene was unrealistic! The family had asked the man to hold their camera. In that case, it is extremely unlikely that he would turn out to be dishonest—in real life, at least. Why? For a simple reason: most people are mostly honest, most of the time. If you choose someone at random, then you will probably be lucky enough to find an honest person who is not a thief and who will guard your item without stealing anything.

If, on the other hand, the man had *offered* to hold the camera, then the scene would have made more sense. In that case, he could well have been an honest, helpful passerby. But he could also quite plausibly be a thief trying to set up a robbery.

This is a situation where a luck trap is involved: *biased observation*. Out of all people everywhere, only a small fraction are thieves. But out of the people who might offer to take your camera, the fraction of thieves is much larger. Why? People who offer their help are either thieves or super-helpful extroverted folks. This much smaller sample of people contains nearly all of the thieves, but contains only a small fraction of all the honest citizens. So among this smaller sample, the fraction of thieves is much higher.[1] If someone offers to watch your belongings, they might be a helpful extrovert, but they might also be a thief, and you might not be lucky enough to only encounter the former.

So in the movie, a simple edit—having the man *offer* to take the camera, instead of the family selecting the man at random—would have made the scene much more plausible. We can learn two lessons from this. First, random luck can be used to help you find honest people and avoid thieves. And second, Hollywood filmmakers should always hire statisticians as script consultants.

This issue doesn't only arise in the movies. A travel writer complained in *National Geographic* of the perils of solo travel, "like lugging all your gear to the airport bathroom because you have no travel companion to watch it for you."[2] But perhaps she was being too hasty. If someone offers to watch her bag, she should indeed be skeptical—perhaps the person is a robber whose offer arose from nefarious motives. But if the writer herself chooses someone *at random* to watch her belongings, she will almost certainly be lucky enough to find an honest citizen for the job.

Now, I am not offering any actual guarantees or promises about this. Indeed, I'm sure that if every person who reads this book asks a stranger to watch their luggage at the airport, or hands their valuable camera to a random passerby, it is indeed possible that someone, somewhere will have a problem anyway. And, despite my best logical, lucky reasoning, I cannot be held responsible if you do. (My lawyers made me put in that last bit. Or at least they would, if I had any lawyers.)

Innocence Lost

Nothing upsets people more than hearing about murders. Killings, especially if they are particularly gruesome or horrific, often lead the news reports and are among the most discussed events. I cannot even imagine the pain of anyone who has lost a loved one in this way.

Among the many terrible casualties of murders is, it seems, the truth. For example, in 2017, President Donald Trump told a group of sheriffs that "the murder rate in our country is the highest it's been in 47 years."[3] This was consistent with the earlier description in his inauguration address of his country's "crime and gangs and drugs that have stolen too many lives," resulting in "American carnage."[4]

The only problem with the president's claim? It was completely false. Indeed, in 2016 the United States had a rate of 5.3 murders per 100,000 people. This represented a small increase from the murder rates in the seven years 2009 to 2015, which were all between 4.4 and 5.0. However, it was still a lower rate than in any of the 43 years between 1966 and 2008, when the rates began at 5.6 and climbed to a high of 10.2 (in 1980) before falling to 5.4.[5] So if he wanted to be negative, he could have truthfully said that the 2016 rate was the highest (barely) in the past eight years. Or, to be more balanced, he could have said that the 2016 rate was the eighth-lowest in the past 51 years. But I guess neither of those true claims would have packed as much punch as the false claim he made instead.

I have some firsthand experience with exaggerated crime rates. At a time when I was a frequent news commentator, my city—generally quite safe—experienced a spike in homicides. That period was immediately labelled as "the summer of the gun."[6] A police detective declared that the city had "lost its innocence,"[7] and one newspaper columnist wrote that "people were tripping over police tape and bullet-riddled bodies on their way to work."[8] It seemed like a very scary time.

How did I deal with media enquiries about this? I looked at the actual numbers! Compared to the previous year, the number of homicides had indeed increased. But by how much? By 25 percent, it turned out, as the rate per 100,000 went from about 2.6 to 3.2.[9] That is significant, and should not be dismissed, and is of course very tragic for each and every victim and their families. But it is not as big a jump as the media made it seem. The new rate was still significantly lower than the 3.9 recorded in 1991,[10] and was still far lower than those found in most US cities[11] and even many other Canadian cities.[12] The final straw was when I discovered that my city's homicide rate that year was still lower than the country's national average,[13] so even in that very year, a random person was safer inside my city than outside it. Needless to say, this was *not* the impression given by the police and media.

(Somewhat similarly, back in 2001, the United States was grappling with media hype about "the summer of the shark," when three Americans were killed by sharks in short order, leading to a surge of shark fears not seen since the release of the movie *Jaws*—despite the fact that the year 2001 actually saw *fewer* shark attacks and deaths worldwide than 2000.)[14]

But can luck really protect us from serious things like murders? Yes, in a way. A murder rate of 5.3 per 100,000 means about one person in 19,000. So for every 19,000 people chosen at random, 18,999 of them will *not* be murder victims this year. Those are pretty good chances. (By contrast, over twice as many Americans die in automobile accidents as by murder,[15] but President Trump has yet to call *that* carnage.) What this means is that, even without taking any special precautions, you and your loved ones will still probably be lucky enough to avoid being murdered. Of course, this does not in any way lessen the tragedy of each homicide itself. But it does provide a certain perspective on the situation, and shows that luck itself provides fairly good protection.

Oh, and what about my city in the years *after* that crime scare? Well, the number of homicides decreased the next year by 12.5 percent and has remained lower in all the years since.[16] It seems the homicide spike wasn't the start of a new trend; rather, it was just really bad luck. And, if the city had "lost its innocence" that year, did it regain its innocence in the years following? That's hard to say—due to the media's *biased observation*, there were virtually no newspaper articles about that.

Blessed Bike

When the weather is decent, I often bike around my city—to get to meetings, to visit people, to get some exercise, to enjoy the lovely lakefront, and more. While biking, I sometimes stop, to enter a building, go for a walk, or make a purchase. Now, my city is generally very safe, but it does suffer a lot of bike thefts.[17] So when I stop, I am always careful to lock my bike to a sign, ring, or pole, to prevent thieves from rushing off with it.

Well, not quite always. Locking and unlocking my bike is a bit of a pain. Occasionally, when I just have to run into a store or building for one minute (to buy a beverage, for example, or get money from an

ATM) and I am feeling *lucky* enough, I will leave my bike unlocked on the sidewalk.

Am I being reckless here? Perhaps. After all, it only takes a second for a thief to grab an unlocked bike and ride away. If my back is turned for even a moment, they could escape with my bike before I even notice. I would have no way of tracking them down, and my bike would be lost forever.

On the other hand, it would be *very bad luck* if a thief happened to spot my bike during the one minute that I was away. After all, there usually aren't *so* many people around when I leave my bike. And we already know that most people are honest and would not steal. Furthermore, my bike isn't super-valuable, so no really sophisticated bike thief would bother to follow me around, looking for a theft opportunity. So, what are the odds that a thief would happen to arrive just as I left my bike unguarded? Pretty small!

Of course, the longer I leave my bike alone, the more likely that a thief will arrive, since the extended time would provide a form of *extra-large target*. Also, the more often I try this particular trick, the more risky it becomes, since that way the thieves get *many tries* to steal it. So I try to restrain myself to only leaving my bike unguarded very occasionally, for very short periods of time. And so far, it has always been there when I returned.

Now, by writing about my habits, I am creating new risk. If my book becomes super-popular, then perhaps bike thieves everywhere will be alerted to my risky ways. They may start to scrutinize my movements more carefully, looking for an opportunity to pounce. If more and more thieves do this, my bike will be more at risk, due to the *many people* trying to steal it. In that case, if I continue to behave in this way, surely one day my bike will get stolen.

On the other hand, if that scenario does come to pass, then surely I will be able to buy a nice new bike—with all of the extra book royalties that I will collect.

Praise the Unlikely

It's no fun telling people how unlikely they are to win a lottery jackpot. Here they are, hoping to win big, dreaming about all that money, thinking about what they'll spend it on, getting excited as the draw approaches. And then I come along, pointing out their extremely low probability of winning and ruining all of their fun.

But there is an upside to my perspective. There are a lot of bad things that are also very unlikely—and that is a good thing!

Take airplanes. I used to be a somewhat nervous flyer. Every time there was a bit of turbulence, I would grip my armrests tightly and look out the window to see if the wing was still attached. But not anymore. Why? Because of probability!

It is true that there are a handful of airplane crashes each year.[18] They can be quite horrific, killing many people, and they usually garner big headlines and extensive news coverage, leading many people to worry. However, there are nearly a billion airplane passengers a year in the United States alone.[19] And nearly all of them arrive safe and sound, without any serious problems. Indeed, it has been estimated that only one commercial flight in five million has an accident involving fatalities.[20] And on any given flight, you only have about one chance in 30 million of dying.[21] Those are pretty good odds.

In short, you will almost certainly be *lucky enough* to survive your next flight, and you really shouldn't worry about that. (If you do want something to worry about, you could worry about arriving *late*, since more than one flight in five is delayed.)[22]

A similar case can be made about other horrible headline-makers, such as home invasions, child abductions, sudden explosions, terrorist attacks, and so on. They are great tragedies. But the reason they make the news is that they happen so rarely. Out of all of the millions of people who *could* have been victims of such terrible events, only a tiny fraction are actually so unlucky. The rest of us will be lucky enough to carry on with our days in peace and security. Unlikeliness is our friend.

While working on this book, I overheard a friend complaining that her hiking partner refused to hike in the beautiful Rocky Mountains. Why? Due to fear of bear attacks! My friend rightly pointed out that bear attacks are so exceedingly rare that they're really not worth worrying about.[23] It seems that even mighty grizzlies are no match for the power of luck.

Nuclear Near-Miss

Our world currently has over 10,000 nuclear weapons, and once had over 64,000[24]—that we know of. Every one of them is pretty scary. But there were two nuclear warheads back in 1961 that might have been the scariest of them all. Millions of lives could have been lost because of them, and those lives were spared only because of—that's right—*good luck*.

It was Tuesday, January 24, 1961. A B-52 bomber flying near Goldsboro in eastern North Carolina developed a fuel leak in its right wing and became unstable. The crew ejected from the craft, which then broke apart and crashed. The only problem? The plane was carrying two nuclear bombs, each with a payload of a few megatons—hundreds of times larger than the bombs that killed so many in Hiroshima and Nagasaki in 1945.[25]

One of the bombs fell by parachute and was found hanging from a tree in a field.[26] It came extremely close to detonating: two experts claim they saw evidence that five of the six safety interlocks had been set off[27] (others say it was actually three of the four safety interlocks[28]), and only one lock remained to stop the bomb from exploding. The other bomb plummeted to Earth without a parachute[29] and was later found in pieces in muddy, quicksand-like conditions. One member of the recovery team recalls that six of the bomb's seven arming steps had already happened.[30] Although some of these details remain unclear, even a nuclear weapons supervisor who strongly disagreed with these accounts still confirmed, in a recently declassified 1969 memo, that "one simple, dynamo-technology low voltage switch stood between the United States and a major catastrophe."[31]

In short, both bombs came very close to detonating and creating huge explosions and radioactive fallout that could have killed millions.[32] Had either of the bombs detonated, this might have been misinterpreted as a foreign attack, triggering a full-scale nuclear war. In each case, most of the safeties had apparently been compromised, and only a single step remained to create an explosion of unimaginable magnitude.

So, where does luck fit in? Well, it was *lucky* that each of the nuclear devices required multiple steps to detonate. Even though many of the steps happened through bad luck, it was still less likely that they *all* would. On the other hand, thinking globally, there are so many thousands of nuclear bombs on the planet, which provides *many tries* for this sort of nuclear accident and thus makes it more likely that some such accident will happen at some time—which isn't very comforting at all.

I guess the bottom line is simply that when both bombs failed to detonate on that fateful day in 1961, we were all *very, very lucky*.

Comforting Luck?

When I write and speak about how luck can protect us and how crime rates are often lower than they are made out to be, not everyone is pleased. For example, one critic wrote, "[Rosenthal] points out fallacies in the oft-heard cry from the media that our crime rate is skyrocketing, by calmly looking at the numbers. But this is where he falters, because quantitative analysis can come across as cold [and] hardly consoles the families of the victims."[33]

When I read this, naturally I wasn't pleased to hear that I had "faltered." But more importantly, this sort of critique has forced me to think about the issues of "cold" analysis and "consoling" victims and so on. Is it the job of quantitative analysis to console? And if so, how?

On one level, the critic's comments seem ridiculous. Of course no statistics can ever truly console the families of victims of serious crimes. If, say, someone's child has been murdered, what statistical analysis could possibly be of any help to them? Whether I point out that crime rates are generally decreasing,[34] or instead make inaccurate claims that the rates are actually rising, the child will still be dead. If anything, surely the best way for quantitative analysis to show respect for victims is to be accurate and truthful in reporting the facts, which is all we can really hope to achieve.

But on another level, the comment struck a chord. When luck hits in seriously negative ways through major crimes, deadly accidents, terrorist attacks, destructive wars, and so on, perhaps formal analysis isn't enough. Perhaps it is indeed important to find ways to comfort or console. But how?

An Unexpected Inscription

In the course of giving talks about probability and randomness, I have had the pleasure of meeting lots of different people and hearing their perspectives. But one meeting stands out in my mind.

I had just given a public talk in a high school gym to nearly 800 people at a wonderful monthly science outreach event.[35] I had discussed the probabilities behind opinion polls, lotteries, homicide rates, casino profits, game strategies, coincidences, and more. Afterwards, I was ushered to the lobby area, where I met enthusiastic readers and exchanged brief words about randomness. Then, my host introduced me to a kindly older couple who were regular attendees.

So far, so good. But I wasn't prepared for what happened next. The host explained to me that the couple's son had recently died of cancer—an unimaginably tragic situation. But the couple quickly explained that, during his last days, their son had taken great comfort from reading, of all things, my writings about probability! They explained that this had allowed their son to see that his cancer was, well, random. It wasn't a punishment. It wasn't his fault. It wasn't because of anything he had done, or not done. It wasn't because he was a bad person, or that he deserved to die. Rather, it was just random (and very, very bad) *luck*.

I was totally blown away by what I was hearing. I had no idea how to talk to parents about a recently deceased child. I wouldn't have thought that discussion of probability would be helpful to them. And yet it was. Randomness had been their friend. Thinking about luck had somehow helped the family to get through an incredibly difficult time. Wow.

The couple even asked me to inscribe a book to their deceased son. This is something I had never done before, and haven't since. I awkwardly wrote a short note to him, thanking him for his wonderful story and his bravery. Although I don't believe in an afterlife, I did feel a strange and special connection to this unfortunate young man and all that he went through. May his memory and courage live on forever.

This story made me realize that there is another way that luck can also protect us. Namely, it can help us to realize when something wasn't our fault. If we can see that our misfortune did *not* have special meaning, then perhaps we can stop blaming ourselves for it. We still have to deal with the consequences of the negative situation, but at least we don't have to deal with any culpability.

I was reminded again of the Serenity Prayer, and its luck version. "Accept the luck I cannot control," indeed. Or, if not accept it, at least recognize it for what is. Cold, stupid, meaningless, cruel, dumb luck.

CHAPTER 12

Statistical Luck

I am a professor of statistics, the science of analyzing data. *Luckily for me*, this topic is now more important than ever. In the modern computer age, there is more and more data available to analyze, from financial interactions to internet usage to medical treatments to demographic trends, and so much more. This may be why, in 2017, the job of statistician was ranked as the best job of all by the website CareerCast,[1] and as the best business job by *U.S. News & World Report*.[2] Loads of companies, from medical researchers to financial institutions to software developers to marketing analysts to computer innovators, are looking for "data scientists"—people who understand statistics and can apply it to diverse types of data. And they are always trying to hire our best statistics students.

Statistics has been saving lives for centuries. One of the earliest examples comes from Florence Nightingale. Concerned about the high mortality rate of British soldiers during the Crimean War in the mid-19th century, she spent two years compiling an 830-page report filled with detailed tables about their causes of death. She developed new ways of presenting the data in chart form to make them more easily understandable.[3] Her studies demonstrated that the number-one reason soldiers were dying was not enemy fire, but rather, poor sanitation in military hospitals, which led to typhoid, cholera, and other preventable diseases. Her advocacy led the British Army to form a special Sanitary Commission to improve hygiene, and death rates were significantly reduced. Nightingale is rightly credited not only as the founder of modern nursing, but also as one of the first and most important applied statisticians. She was elected to the Royal Statistical Society in 1859—the first woman to be so honoured.

So statistics is a wonderful, useful, important, growing area. There's just one little problem: everybody hates it.

My university studies and early research were in mathematics. In my younger days, when people asked me what I did, I would say that I was a mathematician. The most common response to that was "I hated my math class." So I bitterly adjusted to the reality that, even though I love mathematics, my opinion is not shared by all. Then, as my career progressed, I joined a statistics department, and my research interests gradually moved over to more statistical matters. At this point, when people asked me what I did, I would instead say that I was a statistician. And the most common response to that? "I *really* hated my stats class." That's right—I didn't think it was possible, but most people actually hate statistics even more than they hate mathematics. (Sometimes they follow their response with some version of "Oh, but I'm sure I would have liked my stats class more if *you* had been teaching it." But that is too little, too late.)

If statistics is so important, and leads to such good jobs, then why does everybody hate it so much? Well, part of the problem is that statistics classes often get too technical. They get bogged down in esoteric formulas about binomial probabilities and F-distributions and chi-squared tests and normal tables and regression coefficients and hypothesis tests, to the point where the importance and meaning of statistics get lost in the details. Students can't see the forest for the trees, and they struggle to memorize enough techniques to pass the final exam without feeling much appreciation for the subject's value.

It's too bad, though. Because at its heart, statistics is about one single, simple, intuitive, important question: Is it just luck?

Lucky Shot?

The essence of statistics can be boiled down to just one idea. Whenever an effect is detected—say, patients getting healthier after taking some medicine, or an opinion poll saying one candidate is more popular than another—statistics asks one simple question: Is the effect real, or is it just luck? That is, does the data indicate a real difference between taking a drug or not, or between the popularity of one candidate and another? Or, was the observed difference just random luck, signifying nothing?

To put it another way, statistics is all about protecting us from the *lucky shot* luck trap, in which the applicant appears to be a great sharpshooter but actually just got very lucky that one time. Statistics involves analyzing data to determine which aspects indicate true effects and which are just, well, lucky shots.

For example, suppose a certain disease has a 40 percent fatality rate—that is, it kills 40 out of every 100 patients who contract it. Then a pharmaceutical company announces it has a new drug that will reduce fatalities. Suppose that, to test this, the drug is given to 100 patients with the disease. If only 12 of them die, then obviously that is a significant reduction from the expected 40, so the test provides evidence that the drug works well. On the other hand, if 40 or more of the patients die, then the outcome is clearly no better than what would have been expected without the drug, so the test is a failure.

But now suppose that 36 of the patients die. That is a little bit less than the 40 deaths we would have expected without the drug. But only a little. A statistician would then ask: Is this slight decrease "significant," or is it just luck? That is, did the drug actually contribute to a lower fatality rate, or not? To put it differently, if we then give the drug to lots more patients, can we be confident that fewer than 40 percent of them will die? Or is it more likely that the same number will die with or without the drug?

To test this question more formally, a statistician might compute a *p-value*—the probability that we would have seen such a reduction in mortality just by luck, if the drug had no effect whatsoever. In this case, the p-value would equal the probability that 36 or fewer of the 100 patients would have died if the drug did nothing, so each patient had the same unchanged 40 percent chance of dying. In this case, that probability works out to nearly 24 percent. This means that there is nearly one chance in four of achieving such a good mortality result by luck alone. This chance is high enough that most statisticians would say the reduction in mortality was *not* significant. On the other hand, if the p-value were much lower—say, less than 5 percent, or even 1 percent—then the decreased fatality rate would be declared statistically significant—that is, the decrease in deaths was greater than could be expected just by luck alone. (There are also alternative statistical approaches besides p-values, such as Bayesian inference, but they all still boil down to trying to distinguish genuine results from dumb luck.)

So that's statistics in a nutshell. Observe some effect, like a decrease in fatalities upon administering a drug, and then try to figure out if the effect is real or is just luck. And really, how could anyone hate that?

P-value Romance

I sometimes use p-values with my students. In one introductory class, I had the students complete a survey.[4] Among other things, I asked them whether or not they currently had a romantic partner. Of the males in the class, 34 percent said yes. But of the females, only 28 percent did. Aha—this shows that males have more romance than females do, right?

Hold on. Freshman students were the only ones to complete this survey. So perhaps it is only among freshman students that males are more likely than females to be involved in romantic relationships. (The gap could exist because their partners are still in high school, or past first-year university.) Or perhaps the result is even more specific, and it is only male freshman statistics students who date more than the females do. Maybe many of them are dating female literature students, for example. Well, maybe. (Though I don't recall female literature students being so interested in male science students back when I was an undergraduate.)

In fact, the true explanation is much simpler. The result from my student survey was just luck. In terms of luck traps, the higher male romance rate was simply a *lucky shot*. It was just random chance that in this one particular class, a slightly higher percentage of males than females were in a romantic relationship. A different class could easily have had the opposite ratio.

How do I know this? Well, I computed a p-value! A standard statistical test yielded a p-value of 74 percent.[5] This means that even if there were no true difference between male and female dating patterns, survey results as different as those in my class would occur about 74 percent of the time. This is such a high probability that the survey results don't actually prove any true difference at all. Due to the fairly close percentages (34 percent versus 28 percent) and small sample size (only 41 male and 39 female students), the results are not statistically significant. In reality, male and female freshmen may well have the same dating habits, or perhaps the females actually date more. Based on this student survey, no clear conclusion can be drawn.

Well, at least about romance. When it came to heights, a different story emerged. On average, the male students in my class were nearly five inches taller. This time, the standard statistical test gave a p-value of about one chance in 12 million[6]—extremely small. So in this case, it was not just luck: the male height advantage was statistically significant, and thus a true difference. I could thus conclude confidently (if unsurprisingly) that, on average, male university students are taller than female students. But as in real life, romance was a more compli-cated question.

P-value Medicine

The p-value statistical approach is used in medical research all the time to great effect.[7] One doesn't have to look very far to find all sorts of different studies that help improve our understanding of human health by using statistics to distinguish the random luck from the true facts.

A 2016 study investigated why so many children suffer from asthma and allergies.[8] It compared 30 children from an Amish farm-ing community in Indiana with 30 children from a Hutterite farming community in South Dakota. Although the two groups had fairly sim-ilar backgrounds and lifestyles, their asthma profiles were quite differ-ent: none of the Amish children had asthma, while six of the Hutterite children did. This difference was striking enough that the researchers searched for an explanation. They eventually discovered that the dust in houses in the Amish community had much higher levels of endo-toxins, because their barns are closer to their houses and they allow their children to play in the barns near the animals. Could this differ-ent dust explain the different rates of asthma?

They decided to test this hypothesis on mice. They took two sets of 12 mice and treated them exactly the same, except that the second set was also given Amish house dust extracts every two or three days. After 30 days, they measured the resulting levels of a certain asthma-causing antibody, Immunoglobulin E. The first group had average levels of 1,458 (measured in "international units per millilitre"), while the second group averaged just 859. They computed that the p-value for this difference was only 3.1 percent, which was small enough to be significant. This indicated that it probably wasn't just luck—the Amish house dust did indeed reduce the amount of the asthma-causing anti-body. (By contrast, the Hutterite house dust had no such effect.) The study results were published in the prestigious *New England Journal of Medicine*, and may have a significant impact on reducing cases of asthma in the future.

Another example comes from a 2016 study testing a new drug, Dupilumab, for its effectiveness in combatting dermatitis skin diseases.[9] One trial, "SOLO1," randomly placed 224 patients on placebo, and found that 23 of them (10.3 percent) had a significant reduction in their symptoms. A further 224 patients received Dupilumab every second week, of whom 85 (37.9 percent) had a significant reduction. A third group of 223 patients received Dupilumab every week, of whom 83 (37.2 percent) had significant reduction. So, how could these results be interpreted?

Well, not surprisingly, the slight difference between the "every second week" (37.9 percent) and "every week" (37.2 percent) groups was too small to be significant, just like my students' romance difference. So it seems to make no difference whether the treatment is applied every week or every two weeks. But what about the difference between the treatment groups and the placebo group? Was that a significant improvement, or just luck? The researchers ran some statistical analyses and concluded that the p-value for this difference was less than 0.001, or one chance in a thousand. This is small enough to indicate that the results were probably not just luck, but rather, indicated a true improvement from taking the drug. They later replicated their results with a second study, "SOLO2," which came to similar conclusions, making their finding even more convincing. As a result, the United States Food and Drug Administration granted Dupilumab "priority review status" for approval for general use.

A massive study by the Women's Health Initiative considered the risks and benefits of hormone replacement therapy in postmenopausal women.[10] They studied the common treatment of estrogen plus progestin in 8,506 patients, and compared them to 8,102 patients getting a placebo instead. As expected, the patients receiving the therapy had fewer bone fractures. However, in some other outcomes they did worse. For example, 33 (0.39 percent) of the treatment patients died of cardiovascular disease, while only 26 (0.32 percent) of the placebo patients did. Was this difference statistically significant? No. Statistical analysis indicated that this could have been just luck. On the other hand, 694 (8.2 percent) of the treatment patients suffered from overall cardiovascular disease, while only 546 (6.7 percent) of the placebo patients did, and this difference *was* significant (because of the large sample sizes involved), implying that the worse outcomes for the treatment patients were not just luck but were actually attributable to harmful effects of the hormone replacement. Similarly 151 (1.8 percent) of the treatment patients suffered from venous thromboembolism (blood clots in the vein), while only 67 (0.8 percent) of the placebo patients did, and this was also found to be statistically significant—not just luck. As a result, the study concluded that this hormone replacement therapy "should not be initiated or continued" (at least not for the prevention of cardiovascular disease), a finding of potentially great importance.

Of course, that was just one study, and some other studies have reached somewhat different conclusions.[11] And, like everything else, all medical studies are at risk of luck traps, such as *biased observation* due to hidden data, *alternate causes* for some results, researcher biases, problems with replication (more about that later), and more.[12] Nevertheless, these various studies show the power and importance of statistical analysis in resolving complicated medical questions. Without statistics, we could never draw clear conclusions—we could only guess at which treatments are truly helpful and which ones only seem to be helpful due to random luck.

Investor Luck

While I was revising this book, the *Wall Street Journal* published an article appropriately titled "Is Your Stockpicker Lucky or Good?"[13] In it, leading investor Victor Haghani warns that identifying talented fund managers is harder than you might think. To illustrate his point, he imagined two coins: one fair, with a 50 percent probability of coming up heads, and the other biased, with a 60 percent probability of coming up heads. How many times would you have to flip both coins before you could be 95 percent certain which coin was which?

Haghani and his colleagues asked this very question in an online survey. What did they find? Well, half the people thought 40 flips was enough. In fact, a third thought that just ten flips would suffice. However, careful probability calculations reveal the true answer: 143 flips are required to be 95 percent certain of picking the correct coin.[14] That's a lot of flips. In terms of investing, this means you'd better examine fund managers' returns over a very long time period before drawing any clear conclusions about their abilities. For example, if one investor makes the right decision 60 percent of the time, while another investor only gets it right 50 percent of the time, then this advantage will only become clear once they each make 143 separate decisions, which might take an eternity.

The lesson is that if something good happens just a few times—an investor gets a few good returns, or a sports team wins a few games, or a medicine cures a few patients—then this does not necessarily mean that the effect was real, or that the trend will continue. If you don't have enough observations, then whatever success you have observed so far might well be nothing more than just plain, meaningless luck.

Unlucky Climate

There is much talk these days about global warming and climate change. The claim, in a nutshell, is that the planet Earth is significantly warmer now than previously. For example, the Fifth Assessment Report by the Intergovernmental Panel on Climate Change summarizes that "warming of the climate system is unequivocal, and since the 1950s, many of the observed changes are unprecedented over decades to millennia."[15] Al Gore's Climate Reality Project says that "ninety-seven percent of climate scientists agree that man-made climate change is a reality. We know it's happening, and we know why: carbon pollution is warming our planet and creating dirty weather like extreme droughts, flooding, wildfires, and superstorms. And we're all paying the price for it in lives, livelihoods, food and water scarcity, and in every way imaginable."[16] And the US Global Change Research Program writes, "Climate change is happening now. The United States and the world are warming, global sea level is rising, and some types of extreme weather events are becoming more frequent and more severe. These changes have already resulted in a wide range of impacts across every region of the country and many sectors of the economy."[17]

However, some people do not believe that climate change is occurring at all. They think the world's temperatures have remained essentially stable, with any minor ups and downs nothing more than random luck. They dismiss any claims to the contrary as outright lies with ulterior motives. For example, President Donald Trump has tweeted, "The concept of global warming was created by and for the Chinese in order to make US manufacturing non-competitive."[18] Jim Inhofe, a US senator since 1994, has written an entire book about how global warming is nothing more than a hoax.[19] And filmmaker Martin Durkin made an entire documentary describing climate change as "a lie" and "the biggest scam of modern times," one that has been "created by fanatically anti-industrial environmentalists, supported by scientists peddling scare stories to chase funding."[20]

There are many important and complicated questions about climate change, such as what is causing it, how worrisome it is, how it will evolve in the future, and what should or should not be done about it. But what about whether global warming exists at all? Of course, the actual daily weather varies greatly from one day to the next, in ways that are very hard to predict. But long-term trends are different.

For example, NASA has compiled a detailed index of changes in the average annual global surface temperature from 1880 to 2016.[21] Its graph clearly shows that, around 1980, the average global temperature started increasing steadily, to much higher levels than before. Indeed, as they note, the 17 warmest years in this 136-year period were 1998, together with 2001–16, and the three warmest (in increasing order) were 2014, 2015, and 2016.

Still, we can ask, is it just luck? We all know that average temperatures fluctuate somewhat from year to year. Perhaps it is just bad luck that we have experienced such heat in recent years. To test this, we need to determine whether yearly temperature changes are large and clear enough to be statistically significant. And to do that, we need to compute p-values. Using the NASA data, I computed that the average temperature in the 37 years between 1980 and 2016 was 0.74 degrees Celsius (1.33 degrees Fahrenheit) higher than the average temperature in the 37 years between 1880 and 1916. This is quite a large difference. But is it statistically significant? Yes! The corresponding p-value (the probability that this difference would occur by luck alone) is less than one in a million billion, so it certainly isn't just luck.[22] Similarly, on average, the global temperature has been increasing by about 1/140 of a degree Celsius (or 1/78 of a degree Fahrenheit) per year throughout this period, or nearly 1 degree Celsius (1.8 degrees Fahrenheit) in total, and the p-value for that is again less than one in a million billion. These p-values are tiny, and they demonstrate beyond any reasonable doubt that the yearly global temperature increases are indeed highly statistically significant and not just luck.

There is still much to debate about climate change and our planet's future, what actions we should or shouldn't take, and what the future outcomes will be. But the question of whether or not the planet's temperatures have increased in the last 136 years is crystal clear and undeniable. With p-values that small, it isn't just luck. It's real.

Backgammon Cheater?

In the computer age, the game of backgammon has enjoyed a renaissance of sorts, due to the widespread availability of online play.[23] One problem is that this requires the computer program to "roll the dice," at least in the sense of using the computer's built-in pseudorandomness to decide the dice outcomes. And many people do not trust the computer to do so fairly. Indeed, a common complaint on many backgammon discussion forums is that the computer is cheating, by selecting better roll outcomes for itself than for its human opponent.[24]

The situation got so bad that the backgammon software companies had to defend themselves from the accusations of cheating.[25] One company went so far as to publish an article about cheating in its newsletter.[26] And to the company's credit, it argued its case using luck traps! Specifically it considered the claim that the computer was getting more doubles rolls than the player. (Doubles are very advantageous in backgammon, since they allow you to count each die twice.) The company ran a simulation of two million backgammon games, ten played by each of 200,000 simulated users. It found that 6,279 of the players experienced the frustrating situation in which the computer got 60 percent or more of the game's double rolls—much more than the 50 percent it should receive by luck alone.

So, was the computer cheating after all? No. Once again, a luck trap was involved: the trap of *many people*. Those 6,279 frustrated users represented just 3.14 percent of the 200,000 players. Furthermore, each user saw a total of about 76 doubles (by both the user and the computer opponent combined, in all of the user's games taken together). Based on this, the probability that the computer would receive 60 percent or more of those doubles, while the user receives 40 percent or less, is about 4.2 percent.[27] This means that, just by luck, without any cheating at all, we would expect about 4.2 percent of the users to see such a frustrating outcome. So the 3.14 percent who actually had this problem are slightly *less* than the number we would expect on average by luck alone. The computer wasn't getting an excessive number of doubles after all.

This seems to be pretty convincing evidence that the computer was not cheating. On the other hand, the company pointed out that this still leaves over 6,000 users who might "think" the computer cheats. And if just 1 percent of those users post a complaint, that would still generate over 60 accusations of cheating for the company to deal with. That's the problem with drawing conclusions based on just one sample, without considering the *many people* who all have their own samples, some of which might seem much more fair than yours does.

Anecdote Antidote

The best way to see the value of statistical analysis is to compare it to its opposite, the anecdote. People often hear about one case, or a few cases, where some treatment seemed to work or some decision seemed to pay off, and use this to conclude that the treatment or decision must be a good one. It ain't necessarily so!

A major example is provided by "alternative medicine" treatments. This term encompasses many different things, including many sensible techniques to relax and exercise and eat well and feel good. However, it also includes many attempts to circumvent traditional medical treatment for major health problems, in favour of more "natural" therapies.

One common feature of many such treatments is—that's right—anecdotes. Most websites about alternative medicine centres include a page of testimonials, purportedly from real patients, about how the alternative treatment saved their lives and restored their health.[28] Since they are from real patients with real experiences, we can trust them, right? Not necessarily. There may be luck traps at play.

One possible luck trap is *false reporting*. Perhaps those testimonials are completely false, made up by the companies to trick readers into buying their products. But there is actually no need for companies to stoop to such levels. More fundamental traps are available from *biased observation*.

First of all, the company gets to choose which testimonials get seen. Those patients who were not satisfied, achieved poor results, or did not improve their health will not get their reports placed on the company's web page. If reports from *all* patients were available, that would be a lot more valuable.

Even sadder, patients who do not survive do not get a chance to tell others about their experiences. So if a disease is serious enough to be life-threatening, and if the alternative treatments do not help, the patient will die and their health journey may be quickly forgotten. One of many such patients is Michaela Jakubczyk-Eckert, who got breast cancer back in 2001.[29] She initially underwent chemotherapy, which shrunk her tumour significantly. However, she then switched to an alternative German doctor who believed (without evidence) that cancer stemmed from psychic conflict and could be cured by purely psychological means. As a result, she stopped doing her chemotherapy, her cancer became more aggressive, and she died in 2005. As noted by science blogger and doctor David H. Gorski, "Unfortunately, patients like Michaela can't give their testimonials to counter the testimonials of true believers. Never forget that."[30]

A friend of mine had a similar story. She was very energetic and vivacious, going skiing and dancing and running without end. Then, in her 40s, she was diagnosed with breast cancer. Convinced that traditional medicine was dishonest and untrustworthy, she instead turned to alternative therapies. She became convinced that extremely strenuous exercise was the best way to conquer her disease, so she took to that with gusto. I once met her for lunch and was amazed to learn that she had run many miles, for over an hour, to meet me. It was all part of her incredible physical condition and intense exercise regimen. However, despite whatever anecdotes she had heard, breast cancer is not treatable by exercise (though it does have a very high survival rate when caught early and treated by modern medicine[31]). About two years later, her family posted on her Facebook page that she had passed away. I was very sad to learn that this fun, energetic, outgoing person was gone forever. And I couldn't help but think that, if she had only rejected anecdotes and followed the true scientific evidence, she could have received the proper treatment and might have lived decades longer.

So, when considering medical treatment or any other serious matter, don't rely on anecdotes, stories, and testimonials that could be the result of luck alone. Instead, find out about the serious evidence, obtained from proper scientific experiments and interpreted using careful statistical analysis. Base your decisions on what has been clearly demonstrated to be true, not on anecdotes you have heard, or what you hope or wish to be true. It might save your life.

CHAPTER 13

Repeated Luck

A single story or experiment or observation can often be very informative, if we remember our luck traps and interpret the results appropriately. But sometimes, the only way to truly learn is through repetition.

Some years ago, at a vacation resort, I found myself playing water basketball with a friend.[1] In one memorable moment, far from my basket, I had my friend well covered and he was making no progress. In desperation, off balance, without really looking, he heaved the ball far over my head. On it sailed, towards the edge of the pool, before—yes—landing with a big swish right in my basket. Two points for my friend! At the time, we both found this turn of events hilarious, and we doubled over in laughter. Why? Well, because my friend had gotten so *lucky*.

But who's to say? What made my friend's shot the result of luck instead of skill? Perhaps he's actually another Stephen Curry, able to make incredible shots with amazing precision under virtually any conditions. Why are Curry's shots marked down as skill, while my friend's heave instead provoked gales of laughter over how lucky he had been? Or to put it differently, how could we determine precisely whether my friend's feat was an act of luck or skill?

The answer is: *replication*. To evaluate my friend's skill, we should ask him to repeat a similar shot many times. If he scores a basket every time, or even a large fraction of the time, this would demonstrate a high level of skill. By contrast, if he misses the next 20 shots in a row, this would suggest that his first basket had been just lucky and not indicative of any true ability. An intermediate result—say, scoring 5 of the next 20 shots—would suggest that his initial success involved *some* level of skill, combined with some amount of luck.

Similar reasoning applies everywhere. Suppose you arrive in a new town and see a fancy red sports car drive by. This *might* mean that the town is full of fancy cars. Or, it might simply mean that, by luck alone, the first car you saw was a fancy one. The only way to know for sure is to wait for some more cars to drive by and see if the pattern continues. The more repetitions, the more the evidence is convincing.

The James Bond villain Auric Goldfinger understood this principle in a different context when he remarked, "Once is happenstance. Twice is coincidence. The third time, it's enemy action." Apparently Mr. Goldfinger was a closet statistician.

Repeated Gambling

Replication allows us to distinguish the statistical luck from the true trends. One good way to see this is through the age-old activity of gambling.

Consider again the gambling game craps. It involves repeatedly rolling two ordinary six-sided dice and looking at the sum of the two numbers rolled. On the first roll, if the sum is 7 or 11, you win, while if it is 2, 3, or 12, you lose. If it is any other value—5, for instance—you keep on rolling until you either match that 5 again (and thus win), or roll a 7 (and thus lose). These complicated rules were developed for one reason only: they produce a probability of winning of about 49.3 percent, or just under 50 percent.

Does this very slight "house edge" make any difference? Indeed it does—but only over the long run. Suppose, for example, that you repeatedly bet $10 on craps, until you either double your initial fortune or lose all your money. If your initial fortune is just $10, then you will make just one bet, and your probability of success (that is, ending up with $20 in total) is that same 49.3 percent. But if you start with $100, you will make at least ten bets (and probably many more), and your probability of success (in this case, reaching $200 in total before you lose all your money) is about 43 percent—somewhat less than the 49.3 percent on a single bet, but not too bad.

But now suppose you start with $1,000, and again repeatedly bet $10. What is the probability that you reach $2,000 before going broke? Just 5.6 percent, or about one chance in 18. It gets worse from there. If you start with $5,000, your chance of doubling your money before going broke is just one chance in 1.4 million. And if you start with $10,000, it is one chance in ten million billion!

The point is that, as you place more and more bets, that tiny house edge makes all the difference. The random outcomes of each individual roll become less important, and the overall trend—of you losing money on average—becomes the dominant force. With enough bets, the trend becomes less random and more certain—it becomes inevitable that you will lose your money. This phenomenon is the basis of the *law of large numbers*, which says that in a random experiment such as gambling, replication will eventually find a true pattern within all the randomness. If you make just a few bets, anything can happen. But over the long run, you will lose a lot more than you win. And from the casino's point of view, since there are many customers and therefore many bets overall, the house is guaranteed to make a profit in the long run.

For another perspective, suppose you are working security at a casino and you suspect that one of the five gamblers at a roulette wheel is cheating. How do you determine which one? Suppose all five are repeatedly betting on red, meaning that they have 18 chances out of 38 (or 47.4 percent) of winning each bet. Annie is betting $5,000 each time, and she has made 30 bets so far, of which she has won 16 (53 percent) and lost 14, for a net profit of $10,000. Bill is betting $100 each time and has made 100 bets so far, of which he has won 55 (55 percent) and lost 44, for a net profit of $1,200. Carla is betting $100 each time and has made 50 bets so far, of which she has won 29 (58 percent) and lost 21, for a net profit of $800. Debby is betting $100 each time and has made 20 bets so far, of which she has won 13 (65 percent) and lost 7, for a net profit of $600. Finally, Evan is betting $100 each time and has made 2,000 bets so far, of which he has won 1,001 (50.05 percent) and lost 999, for a net profit of $400. That is:

	Annie	Bill	Carla	Debby	Evan
wager	$5,000	$100	$100	$100	$100
bets	30	100	50	20	2,000
wins	16	55	29	13	1,001
losses	14	44	21	7	999
percentage	53%	55%	58%	65%	50.05%
net profit	$10,000	$1,000	$800	$600	$400

Which one do you think is cheating? And how can statistical analysis help you figure it out?

Well, in terms of profit, Annie has won the most—by a very wide margin! However, when it comes to detecting cheaters, the *amounts* of the wagers and winnings are irrelevant, since they do not affect the probability of winning. So we had best ignore the dollar amounts entirely. If we just count the number of bets Annie has placed, ignoring their value, we see she won 16 of 30, which is slightly over half, but not overwhelmingly so. Perhaps she is innocent after all.

By contrast, in terms of winning percentage, Debby is the highest. She has won 65 percent of her bets, despite having just a 47.4 percent probability of success each time. Surely this is suspicious? On the other hand, Debby has only made 20 bets. Winning 13 of them doesn't seem so outlandish. Perhaps she is innocent too.

So, finding the cheater isn't about the money, nor even the percentage of wins. It is all about the numbers of bets. Who won more of their bets—regardless of value—than is plausible by luck alone?

For this, we have to compute the p-value—the probability of winning a given number of roulette bets (or more) just by luck, without any cheating. This can be computed using the *binomial distribution*.[2] It turns out that Annie's gambling wins have a p-value of 31.8 percent, Bill's have 7.7 percent, Carla's have 8.6 percent, Debby's have 8.7 percent, and Evan's have 0.87 percent. What does that tell us? Well, each of the first four gamblers have p-values that are reasonably large—more than 5 percent, anyway—and are therefore not statistically significant, meaning there is no particular evidence that they cheated. Evan, on the other hand, has a p-value of less than 1 percent. This means that, if you play roulette 2,000 times, there is less than one chance in 100 that you will win at least half of the time.

So if anyone is cheating at your casino, it is probably Evan. True, he didn't turn much of a net profit. But to come out ahead even a little, after so many bets, each of which was weighted against him, is already suspicious enough. Such is the power of replication.

In some ways, insurance policies (of all varieties, including property protection, extended warranties, and so on) are kind of like gambling: the odds are against you, and over the long run you will pay more than you receive. Since so many people buy policies, insurance companies benefit from replication, which is why they are so profitable. (This fact was driven home to me when I was invited to speak about the role of randomness in our lives to an annual gathering of insurance brokers. Where did they hold their annual meeting? At a beach resort in Maui!)[3]

Of course, the long run isn't always the whole story. Insurance is used to try to control randomness, by protecting us from the *bad luck* of property damage and broken equipment and lost wages. Sometimes insurance is even a good idea, to protect against catastrophic loss. But remember that the odds are always stacked against you, and all insurance is a bad bet on average. If you do select an insurance policy or warranty option, then the only truly *lucky* party is the company, which is virtually guaranteed to make money through the magic of repetition.

Medical Meditations

The issue of repetition also arises, to some extent, in statistical studies. Suppose a study claims that a certain medicine is effective at treating some disease. Hopefully the study was well designed, carefully executed, and properly analyzed. Probably it obtained a sufficiently small p-value, indicating that its conclusion was statistically significant. But even so, it is still possible that the result arose just by luck alone. (This is all the more possible if the researchers tested many possible hypotheses and medicines before reaching their conclusion, creating a kind of *shotgun effect*.)

So, what to do? The answer is simple: replicate! Get a separate research team to conduct its own study, on its own patients, and draw its own conclusions. If the second team also determines that the medicine is effective, the results become more convincing. If a third or fourth team also joins in, drawing the same conclusion, then the results are nearly certain.

This issue was nicely illustrated by an xkcd cartoon that depicts scientists trying to prove that jelly beans cause acne. They run study after study, giving different colours of jelly beans to different subjects and then measuring their acne levels. They don't find any significant effect with the purple jelly beans, nor brown ones, nor pink ones. But after trying 20 different colours, they find that one colour of jelly beans—the green ones—do appear to have a "significant" association with acne.[4] Of course, in reality this is just a *shotgun effect*: if you run enough experiments, eventually one will yield results that seem significant just by luck alone. But in the cartoon, that doesn't stop tomorrow's newspaper headline from blaring, "Green Jelly Beans Linked to Acne!" Indeed.

Sometimes inaccurate studies are reported, but the situation is rectified later on. One infamous case is the 1998 study by Andrew Wakefield and colleagues that claimed the measles, mumps, and rubella (MMR) vaccination caused autism-like behavioural disorders in children.[5] This set off a wave of anti-vaccine sentiment in the UK and caused many parents to avoid vaccinating their children, which in turn led to new measles outbreaks (the number of confirmed measles cases in England and Wales averaged 163 per year between 1996 and 2005, but averaged 1,061 per year between 2006 and 2009).[6] However, other scientists were unable to replicate Wakefield's results,[7] and sufficient problems were uncovered that the paper was retracted in 2010.[8] The UK General Medical Council investigated, and found that Wakefield was "a dishonest, irresponsible doctor" who was "guilty of serious professional misconduct."[9] The council barred him from practising medicine in the UK. A follow-up editorial in the *British Medical Journal* summarized how Wakefield had "altered numerous facts about the patients' medical histories in order to support his claim" and "sought to exploit the ensuing MMR scare for financial gain" in ways that were not only inaccurate, but also fraudulent.[10] Nevertheless, the anti-vaccine organization Safeminds continued to argue that vaccines were risky, and it funded an additional study on primates.[11] But even that study concluded, "No behavioral changes were observed in the vaccinated animals . . . This study does not support the hypothesis that thimerosal-containing vaccines and/or the MMR vaccine play a role in the etiology of autism."[12]

So, with enough studies and replications, the truth—in this case, that vaccines do *not* cause autism after all—does eventually come out. (Unfortunately some people still continue to claim that vaccines cause autism, even though it is now clear that they do not.)

Occasionally, researchers later refute *their own* findings. One ongoing debate is whether it is healthier to avoid eating any gluten. (This debate excludes those with celiac disease, whose need to avoid gluten is well established.) A 2011 medical research paper said yes, it is.[13] The authors conducted an experiment on 34 subjects and concluded that "patients were significantly worse with gluten within 1 week for overall symptoms, pain, bloating, satisfaction with stool consistency, and tiredness." The paper reported p-values less than 5 percent for each of those symptoms, and the p-value for tiredness was a very small 0.1 percent. Problem solved, right? Not necessarily. Two years later, a follow-up paper came to the opposite conclusion.[14] After 37 subjects were tested, the researchers determined that "gluten-specific gastrointestinal effects were not reproduced . . . [and] we found no evidence of specific or dose-dependent effects of gluten." And the best part? Two of the researchers were among the authors of both papers. Sometimes it is even a challenge to replicate our own work.

Beware of Germs

Replication is a very useful tool, but it does lead to a conundrum: How much evidence should be required before a new result is accepted and taken as fact?

The history of the germ theory provides an interesting case in point. Nowadays, it is common knowledge that diseases ranging from the common cold and flu to chicken pox and smallpox are contagious, and can be transmitted through contact via the associated germs (usually bacteria or viruses), which can be seen under a microscope. This is why we are all encouraged to frequently wash our hands, avoid excessive contact with other people, keep our hands away from our mouths, and so on.

But it was not always so. Before microscopes were powerful enough to see germs, diseases were thought to be acts of God, or the meting out of justice, or purely random occurrences without any cause. What changed our minds?

Some early work in this direction was done by the German-Hungarian doctor Ignaz Semmelweis. At Viennese hospitals, beginning around 1847, he introduced handwashing protocols especially for medical interns performing autopsies.[15] This caused immediate fivefold decreases in deaths from puerperal fever. He and his students soon published papers announcing their results.

Did these new findings lead to better medical practice and less disease? Not right away. Other doctors were skeptical of Semmelweis's claims. Many, subscribing to the ancient Greek theory of *dyscrasia*, believed diseases were caused by an imbalance of bodily fluids, and that bloodletting—not handwashing—was the key to getting cured. Rather than accepting, or (even better) trying to replicate, Semmelweis's results, they dismissed them and continued on as before. Many patients suffered.

About 15 years later, the French biologist Louis Pasteur made his own discovery. His experiments indicated that the reason beverages like beer and milk go bad is because of the growth of certain tiny organisms within them. Furthermore, he found that heating the liquid to a very hot temperature would kill off these organisms and preserve the beverage—a process now familiar to everyone as pasteurization.[16]

Once again, the medical and scientific communities were skeptical. Pasteur's compatriot Félix-Archimède Pouchet insisted that the spoiling of beverages occurred spontaneously after contact with air, without interference from any living organisms.[17] But this time, other scientists paid more attention and resolved to settle the matter. The French Academy of Sciences offered a prize to whoever could shed new light on this controversial issue. In response, Pasteur conducted a new experiment where boiled broths were exposed to air, but only through a filter that prevented contamination by particles. The broths remained pristine until the flasks were broken. The Academy was convinced, and in 1862 Pasteur received the prize, a modest 2,500 francs—worth about US\$28,000 today.[18] Pasteur had numerous further related successes, including the development of several vaccines, and his research has led to safer practices and reduced illness the world over.[19] The Pasteur Institute in Paris is named in his honour.

So, what can this story teach us? Well, when Semmelweis first noticed the decrease in fatalities, it could indeed have been just luck. Doctors were right to react cautiously, at first. Since his theory was counter to the prevailing wisdom, their skepticism was all the more appropriate and understandable. However, as additional experimental results continued to pour in, the medical community should have taken note and sought to study and replicate his findings. That it took so long for proper handwashing practices to catch on is a mark against the medical establishment of the time. Nevertheless, when Pasteur continued to produce new experiments that further supported his theory, the medical community eventually came around, accepted his results, and acted accordingly.

It is appropriate that scientists demanded greater proof before believing in germs. The more different a new theory is from prevailing scientific wisdom, the more cautious everyone should be about accepting it. This is sometimes expressed in the saying, "Extraordinary claims require extraordinary evidence." In Pasteur's time, the allegation that disease was caused by tiny microscopic germs was indeed an extraordinary claim, and scientists were rightly unwilling to change their entire perspective on disease just because one doctor claimed that one experiment demonstrated something new. It is only because Pasteur was able to continue to provide evidence—that is, to *replicate* his earlier results—that scientists were convinced.

Replication Crisis?

Researchers constantly conduct new experiments, trying to find new and interesting facts about medicine, psychology, human behaviour, and so much more. Most of the time, the experiment ends in failure and nothing new is observed. But occasionally the data indicates some effect, and statistical analysis says that the effect is significant—the p-value is less than 5 percent. When this happens, of course, the researchers quickly publish their findings in a high-profile journal.

And yet. The observed data might be due to a true new effect being observed. Or, it might be due to—that's right—pure luck. Perhaps just by random chance, patients receiving a certain treatment happened to get better faster than patients without it, even though the treatment itself had no effect. Could this really happen in a careful, professional study? Absolutely! In fact, the very nature of p-values guarantees that if there is no true effect at all, then a statistically significant result will still be observed 5 percent of the time.

What this means is that some studies—even those published in top-level research journals—might not "really" show anything at all: their observations might have arisen *just by luck*. If this just happened occasionally, then that might not be so serious. But it has been pointed out that if true results are hard to find—which is surely the case in medicine, social science, and more—then researchers will test *many* false hypotheses before discovering true ones.[20] They will then be more likely to obtain a seemingly interesting result by random chance than because it is actually true. The scope of the problem became clear when one study tried retesting 100 psychology experiments and obtained results consistent with the previous experiments in only 47 of them.[21]

On some level, everybody already knows this. It is the reason why medical advice given one year sometimes gets contradicted the next year. To take one example, does the colour of your mucus or phlegm indicate the status of your cold? As I write this, the website

Healthline.com offers a handy chart showing which colours correspond to bronchitis versus pneumonia, rhinitis, and more,[22] and the Cleveland Clinic provides an interactive tool telling us such nuggets as "Green—Your immune system is really fighting back."[23] Meanwhile, a Harvard medical doctor boldly declares that "it has been well established that you cannot rely on the color or consistency of nasal discharge to distinguish viral from bacterial sinus infections, or even whether you're dealing with an infection at all,"[24] and an Australian biomedical professor declares, "We're often told—even by doctors—that green or yellow secretions indicate you're infectious. But this isn't true."[25] Lack of replicability, indeed!

Similarly, research has gone back and forth on the question of whether drinking coffee is good or bad for you.[26] And the studies are very mixed on the question of whether the use of dental floss helps to reduce tooth decay and gum disease.[27] And one big group research project had 29 different research teams each investigate the same question—whether soccer referees give more red cards to dark-skinned players—using the exact same data. The different teams used different methods of statistical analysis and found very different answers, with 20 of them saying yes and the other nine saying no.[28]

Or, for a serious medical example, consider the "liberation therapy" promoted in 2009 by Italian doctor Paolo Zamboni as a cure for the degenerative disease multiple sclerosis.[29] MS patients immediately began travelling around the world and paying thousands of dollars for the procedure, hoping to arrest and even reverse their symptoms and begin life anew. The only problem? Other researchers failed to replicate Zamboni's findings. Eventually enough studies were accumulated to declare a "definitive debunking" of the procedure.[30]

Some have referred to this as a "crisis,"[31] and it has undermined the public's confidence in scientific results. I once got a call from a harried radio news producer asking for advice about how to report medical studies in the face of this lack of replicability.[32] "Do we just ignore modern medicine, and go back to old wives' tales?" he asked without a trace of irony. I had to calmly talk him down from the edge, reminding him that medical science has been hugely successful in improving our lives, despite certain research challenges.

Even from a calmer perspective, the issue remains. If lots of studies have been published, but later found to be false (or at least not reproducible), then what does that say about studies? And what can be done to fix the problem?

Replication Solutions

The above stories illustrate the lack of replicability in scientific studies, which is a serious concern. What should be done in response?

Ideally, researchers should carefully verify and confirm their results before publishing. But this won't always happen. One psychology paper describes an amusing story of a study that seemed to show, intriguingly, that subjects with more extreme political views were worse at judging the shade of grey of text they were shown. In other words, as they put it, "Political extremists perceive the world in black and white figuratively and literally."[33] Before they published, they decided to do one last replication, just to be sure. The second time, they found no effect whatsoever. So they couldn't publish their findings after all. Their immediate reaction, they explained, was, "Why the #&@! did we do a direct replication?" Given the pressure on academics to publish lots of papers, I probably would have had the same reaction myself.

And sometimes, the problem is even worse: when researchers intentionally engage in *p-hacking* by repeatedly adjusting their variables and approaches until they finally find *some* experiment that leads to *some* result they can publish (even though it is probably false). Indeed, the website FiveThirtyEight.com offers an amusing interactive tool that lets you prove almost anything you want about Republican and Democratic politicians and their impact on the economy by choosing which politicians are included in the study (presidents, governors, senators, and representatives), which economic performance measurements are used (employment rate, inflation rate, GDP, stock prices), whether to put more weight on more powerful positions, and whether to exclude recession periods.[34] In each case, trying different options and adjustments leads to a *many tries* luck trap, making it possible to "hack" the results to show a conclusion where none exists.

In response, some statisticians have suggested that the solution is to lower the p-value threshold from 5 percent (one chance in 20) to something much smaller, like 0.5 percent (one chance in 200).[35] Others have suggested that all studies should require independent replication before they can be published, or that different statistical methods like Bayesian inference should be used instead, or other drastic changes to the structure of scientific research.[36] One psychology journal actually went so far as to ban the use of p-values entirely in its pages,[37] a move that generated much controversy among statisticians (some were offended, but some hoped the move would inspire the use of more sophisticated statistical analysis instead).[38] Ultimately, this move did nothing more than sidestep the issue of how to correctly distinguish the true results from the meaningless luck.

But perhaps there is a more direct way forward. For example, the disproval of the MS therapy was very disheartening to MS patients everywhere. But what does it say about the scientific process? Some might call it another "crisis." But I prefer to see it as the system working, eventually, by ultimately discovering that the treatment did not work as originally thought. As a follow-up article in the medical journal *The Lancet* summarized, "Many experts caution against premature promotion of the hypothesis and call for objectivity and scepticism in follow-up studies."[39]

That last suggestion sounds like good advice to me. Indeed, my own solution to the replicability crisis, which may not endear me to all of my statistician colleagues, is to declare that this isn't really a "crisis" after all. Yes, researchers should be encouraged to check more carefully before publishing a claimed "effect," especially when the effect runs counter to usual common sense. And p-hacking should never be permitted. But even so, it is no surprise that many results cannot be replicated. What else would we expect? Medicine and psychology are difficult subjects, full of false leads. Every research paper, even those published in high-ranking journals, should be regarded as merely a *preliminary* assessment, subject to later confirmation or refutation through additional experiments.

If we regard (as we must) all published papers as merely preliminary, then this means that replication or refutation is a very important part of the scientific process. As many have noted, researchers should be required to share their data and computer code to allow others to verify their claims.[40] And even more importantly, the scientific reward system should be adjusted to reflect this reality. Traditionally, scientists receive great credit for discovering something *new*, but very little credit for merely confirming a result that was already published by someone else. This needs to change!

If it is considered a highly valuable scientific contribution to revisit previous results, scientists will manage to either replicate or refute most findings, and thus eventually discover which conclusions are truly replicable and which are not. This will in turn eventually expose the incorrect results from sloppy experiments and insufficient verification and "p-hacking," thus lessening the reputations of sloppy scientists, and ultimately motivating all researchers to be more careful in what they claim.

Problem solved—eventually.

CHAPTER 14

Lottery Luck

I n my media interviews and public talks, I am asked all sorts of questions about randomness and probability, and how they apply to subjects ranging from gambling to sports to polling to medical studies and beyond. But the one topic I am asked about more than any other is lotteries. "What are my chances of winning the jackpot?" "What numbers can I pick to have a good chance of winning a large prize?" "Should I continue to play the same numbers every week?" And so on.

From a mathematician's point of view, the answers to these questions are all pretty simple. (Extremely small. None. It doesn't matter. Etc.) Lottery draws are designed to be completely random and outside of our control, so that we cannot predict or influence them at all. But this simple fact doesn't satisfy the true believers, nor stop some people from trying. Invariably, people (like my friend's girlfriend discussed earlier) ask if there is a way to improve their odds, or if their latest "system" can guarantee victory.

So I spend much of my time answering questions about lotteries, mostly for short interviews on the evening news. Little did I know that one day, lottery questions would lead to a front-page story about corruption and fraud worth millions of dollars.

Lottery Wishes

Most lotteries work by selecting a series of numbers, whether by using balls in an urn or a computer program to generate them. To win (or share) the jackpot, all of the numbers on your ticket have to match the numbers that have been drawn. Now, it is possible that the randomness could be flawed, due to weighted or magnetized balls, or a computer flaw, or some sort of cheating. For example, Eddie Tipton, the information security director of the Multi-State Lottery Association in the United States, was convicted in 2015 of secretly installing a computer program to rig a 2010 draw to win a $14.3 million prize;[1] in 2017 he confessed the details of his scheme, which also extended to lotteries in Colorado, Wisconsin, Kansas, and Oklahoma.[2] But for the most part, lottery draws are conducted properly, so the numbers drawn are equally likely to be any of the available numbers. If so, your probability of hitting the jackpot is simply equal to one chance in all the potential combinations of numbers.

The problem is, there are lots of different potential combinations, so your probability of winning the big prize is very small. For example, consider the US Powerball lottery. Its jackpots have reached over $1.5 billion[3]—a staggering sum that is larger than the annual budgets of most cities.[4] To win (or perhaps share in) the jackpot, "all" you have to do is match five numbers between 1 and 69, plus one additional number (on a red ball, known as the "Powerball") between 1 and 26. Unfortunately, there are a lot of different ways of choosing five balls out of 69—about 11.2 million of them, in fact. Multiplying this figure by the 26 red balls gives a total of just over 292 million possible outcomes. So if you buy a single Powerball ticket, you have one chance in roughly 292 million of winning (or sharing) the jackpot.

Now, one chance in 292 million is really, really, *really* small. By contrast, it is about 28 times more likely that a randomly chosen American will be killed by lightning this year.[5] Or about nine times more likely that a randomly chosen American will one day be president.[6] Or about 42 times more likely that you will die in an automobile accident while driving across town to purchase your lottery ticket.[7] Or equally likely that a randomly chosen woman of child-bearing age will give birth within the next 1.7 seconds.[8] To put it another way: if you buy a single Powerball lottery ticket once a week, on average you will win the jackpot about once every 5.6 million years.[9] Not very often!

But many people are not deterred by these odds. Perhaps they "feel lucky" and think that today there will be some forceful luck that will cause them to win big. Or perhaps some other fortunate occurrence has given them confidence that a win today is "in the cards." They feel they have the power to overcome the odds and win the jackpot. Or perhaps they imagine their life story being written out like a novel, and decide that today is the day they "deserve" to win.

In the Mary Chapin Carpenter song "I Feel Lucky," she wins an $11 million jackpot, has two attractive men competing for her affections, and more. Why does this happen? Because she feels lucky today! She even proclaims that if she flips a coin, she will win either way. No probability rules are going to stop her.

All I can say is *good luck with that.* None of those feelings and emotions will actually change your probability of winning—not one tiny bit. Your chance of hitting the jackpot will remain just as tiny, no matter how you feel.

A clever television news producer once arranged for me to try to "convince" lottery ticket buyers that they wouldn't win.[10] Camera in tow, he had me approach people lined up at a lottery ticket kiosk while holding a small whiteboard, on which I would write formulas about their probability of winning while explaining to them just how unlikely it was. But the buyers weren't interested. They didn't want to hear the odds. They just wanted to buy their tickets and dream big. My equations fell on deaf ears, the cameras rolling all the while.

Science and Jelly Beans

Other people try to take a more "scientific" approach to choosing numbers that are "more likely" to win. For example, one website offers various steps designed to increase your chances of winning, such as "Ensure that your numbers are not all odd nor all even," "The sum of your numbers should be between 121 and 200," and "Refrain from using a Single Digit more than twice."[11] Another person claims to have won millions using a "method" that includes such gems as "Always play the same numbers, because now you've eliminated another set of numbers that probably will never come up again." Indeed, if I had a dime for every piece of advice about how to win millions at the lottery, I would indeed win millions—in dimes alone.

It's easy to see why some people believe such "scientific" theories. After all, it is true that most of the jackpot draws will include a mix of odd and even numbers, and will not have more than two single-digit numbers, and so on. So it's better if your selections have those same properties, right? Wrong!

An analogy might help. Imagine that a pearl has been hidden inside one of 100 jelly beans. You get to choose one jelly bean, and if it contains the pearl, you will be rich. You know nothing about which jelly bean contains the pearl, so you figure that each bean has exactly the same chance—one chance out of 100—of being the special one.

But then you notice that of the 100 jelly beans, 95 are green and just 5 are red. So you decide to be really, really clever by choosing one of the green beans. After all, you figure, there is a 95 percent chance that the pearl is inside a green jelly bean, and you want to get in on the action.

So, were you indeed clever? Was it smart to choose a green bean? Or would it have been better to buck the trend by going red? The answer is: *it does not matter at all*.

Suppose you pick a red bean. The probability that the pearl is in a red jelly bean is just 5 out of 100. But even if it *is* in a red bean, there is still just one chance in five that it is in *your* red bean. So, your overall probability of finding the pearl is 5/100 × 1/5, which equals . . . *one chance in 100*.

What if you pick a green bean? Well, the probability that the pearl is inside a green jelly bean is a full 95 times out of 100. But even if it *is* in a green bean, there is still only one chance in 95 that it is in *your* green bean. So your overall probability of finding the pearl is 95/100 × 1/95, which still equals . . . one chance in 100.

In short, if you pick the more common colour, you have a better chance of getting the colour right, but your chance of finding the pearl is no better and no worse—still just one chance in 100. No matter what colour bean you select, *there is nothing you can do to change your odds*.

And it's the same with lotteries. Indeed, with most lotteries, the probability that the winning numbers will be a mix of odds and evens (not all odd, and not all even) is indeed very high. And, the probability of two or fewer single-digit numbers being drawn is also very high. But it doesn't matter. Whether or not you pick such a combination, you still have exactly the same incredibly low chance of matching the actual jackpot numbers. No matter what "system" you use, *there is nothing you can do to change your odds*.

While writing this section, I came across an author who claims to offer "a way you can improve your chances to win" the Powerball jackpot.[12] And what is his brilliant method? He advises you to select a wide range of numbers: "Spread it out over the spectrum of numbers you have to choose from." Yeah, right, that will win you a big prize. *Good luck with that.*

Smart Lottery Numbers?

So, is there any way at all to be "smart" when picking lottery numbers? Well, sort of. You can't do anything to change your odds of matching the jackpot. But you can change your odds of having to share the jackpot, in the extremely unlikely event that you do win. How? By avoiding numbers that other people tend to choose.

Many people select their lottery numbers using a quick pick feature, where the computer chooses the numbers at random. This is actually a pretty good strategy, since random numbers tend not to be "typical" ones and are thus not extremely likely to be copied by other players. By contrast, an obvious pattern like "2-4-6-8-10-12" might be picked by quite a few players. So if you pick those numbers, you still have the same chance of winning the jackpot. But if you do happen to win the jackpot, you are more likely to have to share it.

Similarly, many people pick lottery numbers based on the *birthdays* of their family and friends. For example, on April 6, 2016, Canada's Lotto 6/49 jackpot numbers were 1, 3, 5, 8, 13, and 31. These numbers are striking, since they are all 31 or lower (and hence could correspond to days of the month), and are mostly 12 or lower (and hence could correspond to months). So the six winning numbers might be chosen by someone combining three birthdays: January 8, March 13, and May 31. Or the three birthdays August 31, May 13, and March 1. And so on.

Did that make a difference? It certainly did. The jackpot on April 6, 2016, was not especially large, so the number of tickets sold was no higher than usual. But whereas often there are no jackpot winners, or perhaps one or two, that day there were *eight* different ticket holders who shared the lottery jackpot.[13] Why? Undoubtedly because many of the winners picked their lottery numbers based on birthdays.

How can you make use of this information? Simple: if you *must* play the lottery, then at the very least, avoid common number choices. Either rely on the computer's quick pick feature, or make your choices without following any obvious patterns. And while you're at it, be sure to include lots of numbers above 31, which don't correspond to any birthdays.

Even if you do select lottery numbers according to this advice, you are still extremely, extremely unlikely to win the jackpot. But just in case you do, at least you probably won't have to share it. *Lucky you.*

The Lottery Retailer Scandal

I have already described all the questions I have fielded about lottery probabilities, how unlikely it is to win a jackpot, how you can't improve your odds by choosing different numbers, how the various "systems" to improve your chances don't actually work, and so on. To be honest, at some point I started getting a bit bored by them. And then, just when I thought I had said everything that could ever be said on the subject, lotteries led me to my most dramatic evaluation of luck ever, involving front-page news items, legislative debate, the firing of two CEOs, several criminal charges, jail time, and payouts totalling over $20 million.[14]

It began rather innocuously. I was contacted about yet another statistical question about some lottery news story. This time it was television producers Harvey Cashore and Linda Guerriero of the CBC's current-affairs show *The Fifth Estate*. By this time I had already done loads of TV news interviews about lotteries, plus I was busy and about to fly to Europe. So I initially declined their request and told them to find some other expert to answer their questions instead. However, luckily for me, they didn't find anyone else. After I returned from my travels, they contacted me again and I finally agreed to hear the details of their story. In the end, I was very glad I did.

The producers' story focused on a soft-spoken elderly gentleman named Bob Edmonds, from the small town of Coboconk, Ontario. Mr. Edmonds always played the same lottery numbers, but (like many players) he left it to the store clerk to check if he had won anything. One day, one of his tickets won a prize of $250,000, but the clerk kept this fact to herself and later claimed the prize as her own. When Mr. Edmonds saw the winning lottery numbers published in a local paper, he quickly realized what had happened. It took him over three years to finally convince the lottery company to pay him (most of) his winnings—on the strict condition that he keep the settlement confidential.

The television producers suspected that the lottery company's insistence on confidentiality was motivated by a desire to hide other, similar cases. So they asked me to investigate this from a statistical point of view. What did the numbers say? Were lottery retailers winning more prizes than expected? And if so, were they stealing customers' prizes, or were they just, well, getting lucky?

To figure this out, we had to determine the total number of lottery ticket sellers, how many tickets those people had purchased, and how many big prizes they had won. Each of these issues turned out to be a challenge. The lottery company knew how many different locations sold tickets, but not how many people worked at them. So we had to estimate that based on a survey. Furthermore, the lottery company had no idea whether lottery retailers bought more or fewer tickets than the average citizen, so we had to estimate that from a survey too. (We concluded that, on average, retailers purchase about 1.5 times as many lottery tickets for themselves as does the average adult; other studies later confirmed that this was pretty accurate.)[15] Finally, while the lottery company had used a questionnaire to *ask* all the major winners if they worked at a company that sold lottery tickets, their records were sloppy and they didn't independently verify the answers.

Nevertheless, by combining the lottery company's questionnaire data with the television producers' survey results and various prize data, I was able to draw some conclusions. I concentrated on "major" lottery prizes of $50,000 and above, within the province of Ontario, during a specific seven-year period. I inferred that lottery retailers had won a total of approximately 200 such prizes. I then calculated, based on all of the information available, that if retailers were playing by the rules, with no cheating, then on average they would expect to win a total of only about 57 major prizes—far fewer. So they were indeed winning more than should be expected. But then the question became: Were their excessive wins the result of cheating by stealing customers' winning tickets, or did they just achieve the equivalent of a *lucky shot*?

Fortunately, this was a question I was able to answer. I knew that I had to compute a p-value corresponding to the probability that the retailers would have won so many prizes just by luck alone. If there was no cheating, the number of major prizes won by retailers should be the result of lots of opportunities: all the lottery tickets bought by all the retailers everywhere. And each of these opportunities has a very small probability of success: the chance of winning a major prize with a single lottery ticket. I knew that random counts like these have probabilities governed by a particular mathematical formula called the *Poisson distribution*. So all I needed to figure out was the probability that a Poisson distribution, with 57 prizes on average, would actually equal 200 or more prizes. That is: What is the probability that the retailers would have won 200 or more prizes purely by luck?

At first, I thought that this probability might be fairly large. After all, 200 is less than four times as large as 57. And everyone knows that lotteries are all about getting lucky, right? So, why couldn't the retailer wins be just the result of good fortune?

Boy, was I wrong. When I did the calculation, it turned out that the probability of the retailers winning 200 or more major prizes was less than one chance in a trillion trillion trillion trillion.[16] This is an unimaginably small probability that would never occur by chance alone. I could safely eliminate a *lucky shot* as an explanation for the retailers winning so often.

Next, I had to check for other possible luck traps. Could there be some sort of *shotgun effect* here, of searching many different possibilities until finally finding one that was surprising? No. If any one individual wins a jackpot, beating odds of millions to one, that is explained by the *many people*—millions of them—who buy lottery tickets. But in this case, we had set out to test just one specific question—whether retailers in Ontario were winning a suspiciously large number of major lottery prizes—and had determined that they were. We didn't consider lots of different groups (besides retailers), or lots of different prizes (besides lotteries), or millions of different individual ticket buyers, or lots of different regions (besides Ontario). No shotgun here.

Was there any *extra-large target* involved? No. Here, the "target" was the major lottery prizes, a clear and fixed and important item. Was there any *placebo effect*? No, the meaning of winning a lottery prize was clear and unambiguous, and not influenced by any psychological perspectives. Was there *false reporting* going on? Well, a little bit: due to poor record-keeping by the lottery company, we didn't know precisely how many retailers there were, how many lottery tickets they bought, or exactly how many lottery prizes they had won. But I wasn't too worried about that, since we had estimated these quantities pretty carefully from surveys and extrapolation, using conservative assumptions where required, and any resulting errors couldn't be large enough to change our main conclusion.

The bottom line was that, no matter how you sliced it, retailers were winning more of the major lottery prizes than could reasonably be expected from luck alone.

Hear Ye, Hear Ye

Based on all of these considerations, I was pretty confident that my analysis of the lottery retailer winnings showed the existence of a significant amount of fraud, and I confidently wrote a report and recorded a television interview saying so. But what I wasn't prepared for was the incredible amount of attention our results received. As soon as the television program was aired,[17] the story of the "lottery retailer scandal" became the nation's top news item. It was on the front page of every Canadian newspaper and the lead story of every television newscast.

That "trillion trillion trillion trillion" figure became a household phrase that I repeated in numerous TV interviews (even one in French).

Reactions were swift and interesting. Lottery company officials tried to refute my analysis, arguing that the Edmonds case was an isolated one. They dared to claim that I had "presented a very simplistic mathematical equation."[18] They even hired their own statistical consultants and paid them to write reports, which questioned my results.[19] Their CEO then had the audacity to assert that my analysis was "missing . . . the frequency of play of the retailers," even though that issue had featured prominently in my report (if not in most of the news coverage).

However, these efforts were drowned out by the overwhelming public outrage—against the lottery company! People apparently feel so attached to their lottery tickets that they took personal affront at the thought that some tickets had been stolen by retailers. Their anger caused opposition politicians to vigorously attack the government in the provincial legislature, citing my name and statistical calculations to argue that the government hadn't adequately protected lottery consumers.[20] The Ontario Ombudsman then investigated, and issued a report[21] strongly attacking the lottery company's lax oversight practices. The government responded by firing the lottery company's CEO (this was the first time my statistical calculations had ever had *that* sort of effect!). It also instituted new procedures, under which the lottery terminals would emit a loud sound of a bell every time a winning ticket is scanned, thus clearly alerting the winner. They brought in a requirement that customers sign their tickets before getting them checked by a store clerk, so as to assert ownership. They even added machines that allow customers to automatically check their own tickets to see if they've won before they hand over their ticket. Overall, it was quite the impact!

And it didn't end there. The police then began investigating certain suspicious cases. I wasn't sure that they would be successful, since the statistics only addressed the issue of *how many* fraud cases there were, not *which* of the wins were fraudulent. But the police kept at it, tracking down "suspicious" wins and asking lots of questions. Finally, they made their first arrest—of a retailer who had taken his customer's ticket and claimed a $5.7 million prize as his own. That retailer pled guilty and served one year in jail. Most dramatically, the four true winners were paid their full prize plus interest, for a total of $6.5 million. That's serious money!

A few other cases also led to arrests. Most interesting, a $12.5 million prize had been claimed by a woman who turned out to be the daughter of a convenience store owner, and who "couldn't remember" where she had purchased her ticket. After several years of investigation, including secret recordings of suspicious-sounding conversations, the police charged the woman, her father, and her brother with fraud. In this case, the police had no idea who the true winners were, so they simply made a public announcement that they were "seeking" the rightful winners. Unsurprisingly, with so much money at stake, they received a "flurry of calls" and had to try to sort them all out. They decided to track every claimant's ticket-purchasing patterns, including the times and locations where they bought tickets, the numbers they usually chose, etc. Eventually, they determined that a group of seven construction workers (and regular lottery ticket buyers) were the true winners and paid them their full prize, plus interest, for a total of $14.8 million. Wow—that is a lot of money arising from one simple statistical calculation! (And no, I didn't get a cut.)

Meanwhile, similar lottery-retailer fraud problems were uncovered in other Canadian jurisdictions, including British Columbia (where they also fired the CEO of their lottery company), Atlantic Canada, and Western Canada. Then, separate media investigations[22] in Arizona, Texas, Florida, Massachusetts, and other US states uncovered related cases in which a few players had won far more lottery prizes than could possibly be explained by luck alone. This implied that in each case there was either additional retailer fraud or various money-laundering and tax-avoidance schemes that involved passing winning tickets to others to avoid scrutiny. Most interesting, the California lottery company instigated a major "sting" operation in which investigators posing as customers asked lottery retailers to check their (fake) winning tickets, while giving the retailers ample privacy to do so.[23] Sure enough, a significant number of the California retailers told these "customers" that their tickets didn't win, before trying to claim the prizes for themselves—and getting arrested in the process.

The problems even extended beyond North America. An elderly lady in central England brought a EuroMillions lottery ticket to her local convenience store. The clerk checked the ticket and told her it was worthless. He even kindly offered to throw it away for her. But what the clerk's display screen had actually shown was that the ticket was worth a cool £1 million (about US$1.4 million). He secretly kept the ticket and later tried to claim the prize as his own. To their credit, the lottery officials became suspicious and investigated. Eventually, the clerk confessed to fraud by false representation, and was sentenced to 30 months in jail, while the woman was belatedly awarded her prize. The National Lottery even introduced a new device to audibly alert customers whenever a winning ticket is checked, in an attempt to prevent similar cases in the future. Oh, and about that clerk? In a delicious irony, his nickname was "Lucky."[24]

Even years later, new cases and new investigations continue to arise. And it all started with tenacious investigative journalism, a bit of statistical analysis, and some careful considerations about luck.

CHAPTER 15

Lucky Me

Most of this book is about how luck affects other people. When discussing others, I can try to objectively evaluate their stories and their claims, and hope to figure out what is true and why.

But what about my own luck? What can I say about that?

I certainly recognize that I had fortunate beginnings. I was lucky enough to be born into a good family, who protected and supported and encouraged me. I lived in a safe country, free of war and famine. I had decent health, and the ability to think and learn. In terms of the Warren Buffett lottery discussed earlier, I certainly drew a good slip of paper and got started off in favourable circumstances. No doubt about that.

But what about my life since then? What happened after I grew up and set out on my own, taking with me whatever advantages I was born and raised with? What did I make of my life, and what role did luck play in my development?

Generally speaking, I think my career trajectory has been fairly successful. I obtained a secure and well-paying job, and I published lots of research papers, which led to some academic awards and recognition. Some of that success was, I think, due to my own hard work, good preparation, perseverance, and natural abilities. So I like to think that I can take some pride in that.

But is that the whole story? Certainly not. As much as I hate to admit it, much of my own success has been the result not of any abilities or clever actions on my part, but of plain, stupid *luck*. Like most people, my success was not at all guaranteed. Embarrassing though it may be, there are numerous times when things could have gone differently, when my life could have moved in different and far less successful directions.

Here are a few examples.

Right Subject, Right Time

Years ago, I started working on my Ph.D. dissertation in pure mathematics at Harvard. My goal was to become a mathematics professor at a nice university somewhere. However, the topics I was studying (mathematical physics, quantum field theory, and string theory) were so complicated that it seemed it would take many years for me to learn enough about them to write successful research papers. Even worse, the topics were so specialized that I wasn't sure if any university would want to hire me anyway. In my first year of graduate studies, I passed all the required exams, and in principle was all set to begin research towards my doctoral thesis. But I felt some despair and frustration, and wondered if I should bother to continue with my thesis at all.

Luckily for me, out of curiosity I had also been auditing a different class, about mathematical probability theory. As a youth I had always enjoyed simple probabilities—things like dice and cards and games. The probability topics in the graduate class were much more complicated than that, but unlike the physics studies, they still seemed to have that same fun and interesting feeling. I soon decided to change my topic, from mathematical physics to probability. After some months of hard work, I made some research progress, which was a great feeling.

My initial Ph.D. research concerned a fairly technical topic—"random walks on compact Lie groups." I completed a couple of research papers about this, which was good. But even though I had switched to the general field of probability, this first part of my thesis research was still pretty specialized. It still wasn't clear that this work would lead to a university position or a successful career.

What happened next was even luckier. My Ph.D. supervisor suggested a new problem that he had heard about, concerning some trendy new computer algorithms called "Markov chain Monte Carlo," or MCMC. So I started working on that. I didn't realize it immediately, but these algorithms were just in the process of becoming hugely popular for many different branches of statistics and scientific research. *As luck would have it*, I found myself in just the right place at just the right time.

My mathematical work on these special algorithms appeared at just the right time, so it ended up being widely cited in other research papers. This in turn helped to improve my research stature and reputation, and eventually led to nice academic job offers from a number of different universities. I was off to the races. I would have a successful academic career as a professor. *Lucky me.*

And if I hadn't switched my topic to MCMC when I did? I might well have had a less successful career, doing less relevant research at a lower-level university. Or I might have failed to become a professor at all, and might be working as a low-level computer programmer instead. I can't say for sure. But it seems likely that my research success was based largely on the happenstance of switching to the right topic just when I needed to. In short, on luck.

Simple Percentages, You Say?

I have already discussed how lucky I was to get to do the statistical analysis for the lottery-retailer scandal, which garnered so much publicity and led me to many other opportunities. And that I was even luckier when, after I initially declined to get involved, the television producers couldn't find a substitute so they decided to invite me again.

But I was lucky in another way too. After the publicity that resulted from the lottery story, I was invited to give a 90-minute address to a large conference of police fraud investigators.[1] I was a bit nervous about how to present the story and the related statistical analysis in a way that would be interesting and entertaining for the hundreds of police officers, government officials, and policy-makers who would be present. But I prepared hard, created interesting slides, and developed various examples to aid in the presentation.

And in the end it went great. The audience laughed at my jokes, took in the story, thought about the probabilities, and responded to my questions, and my presentation was a big success. *Phew!*

However, there was one moment when it almost fell apart. I had created a probability scenario and asked the audience to try to guess which answer was which. While waiting for them to raise their hands, I got an uncomfortable feeling. One of the numbers on my slide didn't seem right to me. And if it was incorrect, that would throw off my entire point.

I made a quick decision to simply ignore my suspicions and continue the presentation without change. It all went well, and no one noticed my error or raised any objection.

Several hours later, in the privacy of my own home, I checked my calculation again. Sure enough, I had made a stupid mistake. I had, correctly, worked out one of the probabilities to be something like 0.30245. Then, for my presentation, I needed to convert this value to a simple percentage. Working quickly, and thinking that this was the easy part, I simply wrote that it was equal to about 3 percent. Wrong. As every elementary school student knows, this value actually equals about 30 percent, not 3 percent—big difference. And this mistake meant the point I had made in my talk was, well, not completely wrong, but not completely correct either. Oops!

Luckily for me, no one had picked up on my mistake, and they all thought my talk was great. And while I am usually a very honest person, this time I decided that honesty was *not* the best policy. I quickly edited my slide to correct the error. And I never breathed a word to anyone about the mistake I had made.

Well, until now.

Screech!

My earliest luck-related memory dates back to when I was quite young, probably about six or seven years old. We lived in a safe, quiet neighbourhood, and one day I went walking on some sort of errand with my older brother. Our journey took us across a nearby major street. My memory is fuzzy, but I believe my brother went hurrying ahead to cross the street just before the traffic light changed to red.

The only problem was, I was following behind him and struggling to keep up. I wasn't used to wide streets and fast cars and traffic lights. I chased after my brother without regard to my surroundings. And that's when the sound came.

Screeeeeeeech! A car somewhere was jerking to a stop. Then a man's voice was shouting at me. "You tryin' to get yourself killed?" he demanded. I jumped back onto the divider in the middle of the street, confused and disoriented but now out of harm's way.

At some point the car must have driven off. But I didn't see it. I never quite figured out where the voice had come from. Nor do I really know how close the car had come to hitting me, nor how far out in the street I was, nor when the light had changed.

But I do know what *could* have happened. If I had been a little farther out, or the car had been a little closer, or its speed had been a little greater, I could have been hit. I could have been injured. I could have broken bones. I could have been put into a coma. I could perhaps have died. I was, unquestionably, extremely *lucky* to escape with nothing more than my pride and confidence damaged.

I do remember what happened next. I may have been frightened, but I still had my priorities straight. My first words to my brother were a request, which he readily understood and granted: "Don't tell Mom."

Deal or No Deal?

I receive all sorts of media enquiries related to probabilities and randomness. Usually I am able to handle the questions well enough, without too many big surprises, so *luckily* it usually works out okay. But one time, I had an unexpected challenge.

The topic was the television game show *Deal or No Deal* starring Howie Mandel. On that show, each contestant gets a briefcase with an unknown amount of money in it. As the show progresses, the contestant repeatedly gets offers of varying amounts of money to surrender their case. Each time, they have to decide whether to accept the money offered, or decline in the hope of getting more money later on. A newspaper reporter started off by asking me for my general advice about what decisions the contestants should make, in the context of the show's rules.

I started by giving some general advice that the best decisions are those that will increase your money *on average*. So contestants should compare the offer they receive to the average of the amounts in all of the remaining briefcases, and use that to decide whether or not to accept the deal. I even made some connections to the related principle of *utility functions*, and how you should think not just about how much money you might win, but about how valuable that money would actually be to you personally. All in all, it was a pretty good response, and I was feeling confident that I had conducted another successful media interview.

Then the reporter threw in a twist. He presented me with four different scenarios that had actually occurred on the show. In each case, he described for me the offer, and the amounts in all of the remaining briefcases, and then asked me what decision I would have made if I had been the contestant. He was putting me on the spot!

Nervously, I considered each scenario, computed averages, and reasoned it out. Then I bravely announced my decisions to the reporter. He carefully wrote them down, said goodbye, and walked out of my office. What had I just done? Would I end up looking like a fool?

The newspaper story wasn't published until 12 long days later. Sure enough, it included details of all four scenarios. *Gulp!*

So, how had I done? Well, in one of the scenarios, the contestant and I had each made the same decision, so the newspaper had rightly written, "Who played better: Tie." But in the other three scenarios, my decision was the opposite of what the contestant had done. Game on! And the verdict?

Incredibly, in all three of those scenarios, my decision turned out to be the better choice. In all three cases, the newspaper wrote, "Who played better: Rosenthal." I was victorious! The article was the lead story in the newspaper's Arts section, with the headline "If you're a guest on Howie Mandel's show, you should bring Jeffrey Rosenthal," advice to "leave the parents and their platitudes at home and bring along a mathematician," and full details of all four scenarios.[2]

It was all great publicity for mathematics, for probability, and for me. (My publisher even presented me with a framed copy of the article.) But here's the thing: my success in those scenarios was only a little bit because of my wise ideas about taking averages. It could easily have gone the other way instead. Mostly, I had gotten very *lucky* to emerge victorious in so many cases.

Funny Music

I took piano lessons as a child and later learned guitar too. As I grew, I did a bit of performing and had dreams of making it big and becoming a rock star. Unfortunately, I soon learned just how many other people had the same dream, and how unspecial my own music was. Sure, I could entertain a group of friends, or perform acceptably at a small club, or even regale crowds of drunken statisticians at a conference banquet, but that was about as far as I would ever go.

Some years later, I started doing improvisational comedy. I wasn't a natural, but after a few years of workshops I got to the point where I could perform in scenes, made up entirely on the spot, and mostly do a decent job of keeping the plot going and the characters interesting while getting some laughs. Improv even had a bit of a probability connection, since the random developments on stage kind of resembled the randomness I was researching at university. This was all very exciting, and the skills I developed even helped me to become a better teacher and public speaker. However, it became clear that there were nearly as many people trying to become comedy stars as music stars, and once again I was nothing special.

But then, one day at an improv show, I heard that the performers were looking for a musical accompanist. Someone to sit near the comedians and use a piano keyboard to provide whatever music and sounds would help to improve the performance—happy music when the scene was joyous, scary music when a menace appeared, organ sounds to accompany an imaginary sporting event, and so on. Their usual accompanist wasn't available. Could anyone help them out?

I had never before provided musical accompaniment for a comedy scene. But *luckily*, I had most of the essential skills. I had performed improvisational comedy and had some feel for it. I knew how to play different kinds of music. I had even done a fair bit of musical jamming with friends, situations where I had to adjust my music quickly to fit in with the other musicians. *As luck would have it*, I was again in the right place at the right time. I performed musical accompaniment at the show that night, and did pretty well. The other performers appreciated my efforts. I was in.

After that, it was smooth sailing. Despite the thousands of aspiring musicians and thousands of aspiring improv comedians out there, only a few are able to combine the two arts together into improv musical accompaniment. So after years of being mostly ignored as a musician, and mostly ignored as an improv comedian, suddenly I was in demand as a musical improviser.[3] Producers started calling me to perform in their shows—and *paying* me to do so! Experienced successful improvisers, who once wouldn't have given me the time of day, now made a point of asking me to accompany them. I got to be an onstage musician in a full-length play.[4] It certainly wasn't the big time—mostly just small shows in small clubs—but it added up to quite a lot of performances. I wasn't super-talented, but I had become specialized enough to be needed. Indeed, the week before I wrote this paragraph, I performed at a big New Year's Eve variety show in a large theatre[5] and had a great time.

In my honest moments, I have to admit that these opportunities weren't entirely due to actual merit. I am just a middling musician and a middling improviser. Sure, I can entertain a room full of friendly people. But my individual abilities just aren't strong enough for a producer to bother calling me to perform music or improv at a show, when there are a thousand others with more talent. However, *luckily for me*, by combining these two skills, I found a way to offer something that many shows needed, but few performers could provide. And so, by a stroke of *good luck*, I have had much more of a performing career than I could ever have expected.

Juggling Fruit

After I was an established professor, I went to a talk on campus by Marla Shapiro, a medical doctor and television personality. After her talk there was the usual question period. But none of the 500 people in attendance stepped up to ask anything. After a few awkward moments, I decided to save the day by walking to the microphone and asking a question myself.

In her talk, Dr. Shapiro had discussed issues of work-life balance and how important it is to take some time away from your career to spend it relaxing with family and friends. So I asked her if she herself, as a doctor and media star, found time to do that, and how. It seemed like a reasonable question, good enough to end the silence and give her the opportunity to discuss some further thoughts.

But what I didn't know is that Dr. Shapiro was prepared for such a question. She quickly picked up three pieces of fruit that she had hidden behind the podium for this precise purpose, and handed them all to me. Balancing your life demands, she explained to the assembled masses, is like trying to keep multiple objects in the air. To illustrate the point, she asked me to try to juggle the three pieces of fruit, right in front of this huge audience!

Clearly, her plan was that I would fumble and drop all the fruit, looking like a fool, thus allowing her to illustrate the challenges of balancing many different demands all together. But unbeknownst to her, I had practised juggling a little bit as a youth. I was good enough that I could *usually* keep three tennis balls going, and only occasionally drop one. So that gave me a bit of a fighting chance.

But *fruit*? That was different from tennis balls! Especially in this case: the fruit consisted of one orange (nice and round), one green apple (also pretty round), and one banana. Now, bananas are not round. Not at all. They look and feel and move completely differently from apples and oranges. They are difficult to catch and throw. And I had never tried to juggle one before.

Somehow, ignoring the hundreds of onlookers, I decided to give it my best shot. I put the banana and apple in my right hand, the orange in my left, briefly tested their weight, and then tossed the apple into the air. Just before it came down, I tossed the orange from my left hand, and then caught the apple. So far, so good. But the banana test was still to come.

As the orange descended towards my right hand, I pushed the banana upwards. I managed to catch the orange with my right hand, toss the apple with my left hand, and then catch the banana in my left hand. It worked. I was actually doing it!

As I continued, my confidence grew. The audience vanished. Dr. Shapiro was gone. There was just me, the awkward professor, alone with three objects. Carefully throwing and catching, concentrating completely, I continued juggling the fruits, nice and steady, without fault.[6] Finally, after about ten seconds and many throws, I managed to catch all three fruits in my hands again. I was done. It was over. I had won.

The audience burst into applause. I was the hero. Dr. Shapiro still tried to make her point, but she was reduced to saying that balancing demands is like the way *most* people would try to juggle fruit, "perhaps not so ably as this gentleman here" (meaning me). I felt great.

After the talk, several people asked me if it was all a set-up—if Dr. Shapiro actually knew all along that I would be able to juggle the fruit. Not at all, I explained. Dr. Shapiro didn't know me at all. And besides, my successful juggling was actually of no help to the point she was trying to make. It hadn't been a set-up at all.

Rather, the one time I was asked to juggle in front of hundreds of people, it so happened that I had a little bit of juggling ability and managed to throw the fruits just right, gained confidence, and juggled better than I ever had before. In short, once again, I had just been *very, very lucky*.

Pedestrians Unite

Around the time the lottery scandal finally quieted down, I was contacted by newspapers about a different issue: pedestrian deaths. In a single month, 14 different pedestrians had been struck dead by automobiles in the Toronto area—far more than usual. Initial news headlines were apocalyptic. Surely this was the start of a new trend, and walking in the city was more dangerous than ever before—right?

After my work on the lottery scandal, the media expected me to raise similar alarms about pedestrians. But I tried to calmly consider the issues. Were there any luck traps involved?

It turned out that there were.

First of all, this count of 14 deaths was for the entire municipal region; only seven had taken place within the actual city boundaries. It seemed unfair to combine different cities' fatalities together in this analysis, and anyway, it was hard to do fair comparisons over such a large region, owing in part to insufficient data. Right away, I decided to focus on just the seven deaths within the city itself.

The next question was: Compared to what? How many pedestrian deaths would be considered normal or usual for the city? It turned out that, over a longer period, the city averaged about 32 pedestrian deaths per year, or about 2.66 each month. Compared to that, seven was more than usual, but not *so* much more than usual.

But the real issue was that there was a form of *shotgun effect*. In this case, the different "bullets" corresponded to different months. Pedestrian deaths had been monitored for years, and it was only in this one month that the count was high—if it had been as high the previous month or the month before that, we would have heard about it. And it wasn't so surprising that *some* month would have such a high count.

Indeed, I could compute this more precisely, using a probability method called the *Poisson distribution*. It told me that, in the absence of any new force or special meaning, each month had about a 1.9 percent chance of having seven or more pedestrian deaths. That's pretty unlikely. But it did mean that we would expect such a large count about once every 4.4 years, just by chance alone. Indeed, the phenomenon of getting high counts every now and then, just by chance alone, is called *Poisson clumping*.

So I boldly proclaimed that the high pedestrian death count was the result of chance alone, and not particularly surprising, and the count would probably return to more "normal" levels in the months ahead. My analysis was duly noted, with newspaper headlines like "Not so rare for rarities to occur in waves: Professor," "Maybe there is no meaning behind the numbers, just probability," and "January pedestrian deaths a statistical hiccup: mathematician."[7] These articles were great publicity for me and for statistics. More importantly, they helped to set the record straight and avoid unnecessary panic. So far, so good.

But then I started to get nervous. Perhaps I was wrong. Perhaps this really *was* the start of a new trend. Perhaps the pedestrian death count would keep going higher and higher, sending a wave of new fear through the city. And perhaps they would then look back with scorn on the foolish professor who thought it was just a random blip. I could be ruined!

So I was anxious to hear follow-up statistics, to see how the pedestrian death count changed the next month. Unfortunately, for the longest time, I couldn't find any clear information. Finally, about two months later, I managed to get figures for the month and a half following the month in question. Over that period, the expected number of pedestrian deaths in the city would be about four. So, did the actual count greatly exceed that figure, indicating continuing danger?

On the contrary. The count for that period was only two—somewhat *fewer* than the usual average. So I was right—that high count had been just a statistical blip consistent with Poisson clumping, and nothing more, with no indication of any actual increased danger. *Luckily for me*, what I had told the newspapers was indeed correct, so I wouldn't end up looking stupid. *Phew!*

The only problem? No newspaper published any story about the count returning to normal. Not a single one. I guess when the death counts aren't high, the media quickly loses interest. *Unluckily*, I had to celebrate my statistical victory all by myself.

Nevertheless, I was reminded again about the importance of luck in my advancement. From stumbling upon the right research topic, to mistakes that nobody noticed, to predictions that just happened to come true, my career success has been about a lot more than ability and hard work. Numerous times, luck has also played a crucial role.

CHAPTER 16

Lucky Sports

Professional sports, of one kind or another, are extremely popular in virtually every country. People love to watch, to cheer, to hope, to dream, to bet. As we have seen, fans sometimes ascribe sports outcomes to supernatural phenomena such as the Curse of the Bambino. But on a more serious note, they also like to try to make careful predictions about who will win and who will lose.

Every now and then, I am contacted by members of the media asking for the probability that a team will win a game, make the playoffs, or win a championship. Some people might base such predictions on lucky charms or other superstitions, but I always try to base them, as best as I can, on actual probabilities.

Of course, probabilities can only take you so far in sports. In the end, every game comes down to the actual performances on the field, and which team performs better, and, well, whatever *luck factors* might turn out to be involved. Statisticians can't predict the outcome of games with any sort of certainty, nor even really say what the probabilities are. But that doesn't mean we can't try.

March Statistical Madness
My biggest look at sports luck concerned the NCAA Division I Men's

Basketball Tournament, commonly known as March Madness. This tournament features 64 different US college teams (after four more teams are eliminated in a preliminary "play-in" round), each vying for the championship. A total of 63 games are played over the course of about three weeks, with each losing team eliminated until, finally, just one team remains.

Almost as big as the tournament itself is the hype around bracketology, the effort to predict the complete tournament bracket by identifying the winner of each of the 63 games. Incredibly, the billionaire Warren Buffett once offered \$1 billion[1]—that's right, a billion with a B—to anyone who correctly predicted the entire bracket. (It was even helpfully pointed out that a billion dollars, stacked in one-dollar bills, would be about 68 miles high.)

Was Buffett crazy? Not entirely. Since each game could be won by either of two teams, the total number of different possible tournament outcomes (or *brackets*) is equal to 63 different twos all multiplied together. This works out to over nine billion billion, an unimaginably huge number. Suffice it to say that, to get the entire bracket correct, you would have to be *very, very lucky*. Indeed, no one won Buffett's billion, nor has anyone *ever* managed to correctly predict the complete tournament bracket.

But that doesn't stop people from trying. Each year, there are numerous contests and pools and competitions in which people try to forecast the bracket as accurately as possible. Most predictions use some combination of sports knowledge, study of the players involved, and gut feeling. In March 2013, looking for a different take, some TV producers came calling with a request for a purely statistical approach to bracket prediction, one that didn't make use of any actual sports knowledge or understanding.

I initially hesitated, since I was quite busy at the time, and I could see that a lot of effort would be required to track down various statistical measures and come up with a decent prediction. Then the producers offered to pay me a thousand bucks, so I got right to work.

The biggest challenge was organizing the various data. It turned out that lots of facts and figures about the college basketball teams were freely available online. But each web page had different details, took different approaches, used different abbreviations for the team names, and so on. Much of my time was spent just compiling and coordinating the various numbers into formats that my computer programs could read correctly.

Once *that* was all done, I needed to come up with a method of predicting. I decided to take the previous three seasons' worth of data. I put together a computer file that specified, for each team in each regular (pre-tournament) season, what fraction of games they had won, how they did in their final three games, how "efficient" their offence and defence had been, how strong their opponents had been, and so on. I then examined the relationship between each team's regular-season performance statistics in a given year and how they did in that year's March Madness tournament. I used a simple technique called *linear regression* to combine all of this information together and determine how to predict, as best as possible, a team's result in the tournament.

Eventually I had my final formula. The TV people wanted a catchy name, so I called it the "Rosenthal Fit." Testing my method on the previous seasons, I saw that it predicted the correct winner in about 72 percent of the games, slightly better than the tournament seed rankings (which got about 70 percent correct) and the famous Ratings Percentage Index (which got about 67 percent correct). I was ready to go.

I then applied my formula to the 2013 season. My calculations gave me a single number for each team, corresponding to my statistical model's rating of how well they would do in the tournament. Duke scored highest at 24.1150, Louisville was next at 23.7559, and down it went, all the way to poor Grambling State University at just 3.0220.

From there, it was easy to fill in my bracket. For each upcoming tournament game, I simply had to look at the two teams' scores according to the Rosenthal Fit. Whichever team had a higher score, I would predict to win that game. No problem.

I wrote a little article about my method and scores, and it was published on the TV station's website.[2] I also recorded an interview, which was broadcast on the air, about my methods. And then I just had to wait! I had no idea how my statistical rankings would do. But I did know that *if* my predictions worked really well, the TV station would interview me again, and fame and fortune would surely follow.

Finally the tournament got underway. The early results were mostly unsurprising, since in the first round the strongest teams play the weakest teams. My statistical model was hanging in there, but the real tests were yet to come.

Then, finally, there was a major upset: 12th-seeded Oregon beat fifth-seeded Oklahoma State, surprising almost everyone. And incredibly, my statistical model had predicted that![3] The Rosenthal Fit values were 21.9407 for Oregon, and just 21.5885 for Oklahoma State, so I had anticipated that Oregon would win. Wow! For the first time, I started to wonder if my statistical model might be showing great insights. Maybe I *would* become rich and famous as a statistical basketball forecaster, after all.

Alas, my celebrations were short-lived. There was soon another upset when New Mexico lost in the first round.[4] They were seeded third, and thus had been expected by everyone to do pretty well. And my statistical model had ranked them very highly, at 23.5325, strong enough that I had predicted them to win their first four games. Once they were eliminated in the first round, my entire bracket started to look a lot worse. I quickly realized that my prediction of the

Oregon upset had been, well, just a *lucky shot*. Other teams with high Rosenthal Fit values started to lose too, and my predictions ended up being rather mediocre. Fame and fortune had eluded me. The television people even stopped replying to my emails.

And what was the single worst part of this whole story? When New Mexico lost in the first round, thus ruining my predictions once and for all, they were beaten by Harvard. That's right, Harvard—the very school where I received my Ph.D.! Now, Harvard hadn't even *qualified* for the March Madness tournament between 1947 and 2011, and they had never before won any game at the tournament.[5] But no, they just had to pick this one particular moment to finally win a game, and thus destroy my March Madness bracket. Thanks a lot, Harvard.

Play Ball!

My city's baseball team, the Toronto Blue Jays, won the World Series championship in 1992 and again in 1993. But they haven't done as well since then. Because of this, each time they perform at anything close to championship level, the entire city gets excited, and inevitably I get a call asking for the odds.

And so it happened that a television producer phoned me on September 22, 2015. The Jays were leading their division, just a few games ahead of the dreaded New York Yankees. They had 12 games left to play in the regular season—including one against the Yankees that very evening. So the usual questions came: Would the Jays win that night? Would they win their division? Would they win the World Series? What were the odds?

I took those questions one at a time. That night's game was actually the trickiest thing to predict, since any single baseball game features so many possible variables. I didn't have time for a detailed analysis, so I kept it simple. Up to that point in the season, the Jays had won 57.33 percent of their games, the Yankees 55.03 percent. I assumed that their winning potentials followed the same ratio—in other words, that the Jays had a probability of 57.33/(57.33 + 55.03), or about 51 percent, of winning that evening—and a 49 percent chance of losing.

For the question of winning the division, I made some simple assumptions about the chance of each of the two teams winning each of their remaining games. Based on that, I computed the probability that the number of Jays victories *minus* the number of Yankees victories would be large enough that the Jays would still be ahead when the season ended. My computations gave the Jays an 88.5 percent chance of winning their division.

The World Series question was also challenging. If the Jays won their division, they would enter the playoffs and be one of eight teams vying to win it all. Those eight teams would all have similarly impressive win ratios, and without a lot of careful sports analysis, it was hard to distinguish their chances. So for simplicity I assumed that each team would have about the same chance of becoming World Series champions—about one chance in eight, or 12.5 percent. Which also meant that, unfortunately, the Jays would have an 87.5 percent chance of *not* winning the World Series.

My predictions aired on the television news that evening for everyone to see[6] and got a fair bit of attention, comments, and retweets. At that point, I could only wait to see what outcome fate—or at least the baseball players—would bring.

So, what happened? Well, the Jays lost 6–4 that night.[7] A couple of weeks later, the Jays hung on to win their division, six games ahead of the Yankees.[8] In the playoffs, they won the division series against the Texas Rangers, but then sadly lost the American League Championship Series against the Kansas City Royals. Alas, there was no baseball championship for us. (On the positive side, the Royals themselves went on to win the World Series. So, since the Jays lost to them, then that is kind of like coming in second place, right?)

In terms of my predicted probabilities, the Jays lost that evening's game (probability 49 percent), won their division (probability 88.5 percent), and failed to win the World Series (probability 87.5 percent). Those are all fairly high probabilities, considering. So I guess my predictions were mostly pretty good. Considering.

Go Leafs Go!

My city has popular teams in many professional sports—not just baseball, but also basketball, football, soccer, and more. But in Canada no sport is more important than hockey, and there is no hockey team more popular than our beloved Toronto Maple Leafs. As I mentioned, the last time the Leafs won the Stanley Cup was five months before I was born, and every new hockey season since then has only produced more frustration.

On April 12, 2006, I got a call from a local newspaper columnist. Once again, things were looking bleak. In that season, up to that date, the Leafs had won 38 games and lost 40 (eight of those in overtime), with four games left to play.[9] Would they make it to the playoffs? Only the top eight teams in their conference would advance. Six teams were so far ahead of the Leafs as to be uncatchable, while three others were slightly ahead. To snag the last playoff spot, the Leafs would have to finish ahead of at least two of those three teams. Could they do it? What were the odds?

I checked out the upcoming schedules and starting calculating. It quickly became clear that if the Leafs edged out Tampa Bay, they would almost certainly beat Atlanta as well, and thus make the playoffs. But if not, they had virtually no chance. So essentially, everything came down to whether or not the Leafs could catch up to Tampa Bay, who at that point had 42 wins and 32 losses (five in overtime). The league awarded two points for each win, plus one point for each loss in overtime, which meant Tampa Bay had 89 points to the Leafs' 84. With only a few games remaining, the Leafs would have to win about three more games than Tampa Bay to make the playoffs.

Could they do it? Probably not. Using some simple assumptions, I computed the probability of the Leafs ending up ahead of or tied with Tampa Bay in the regular-season standings to be about 5.8 percent, or just under one chance in 17. Those are pretty low odds. I duly sent my calculations off to the newspaper columnist, aware that I was once again the bearer of bad news. His column ran the next day, including my calculation and discussion, along with my explanation that fans often overestimate their team's chances due to *biased observation*—remembering previous successes more than failures.[10] I hated to bring the city down, but I had to report the probabilities and luck traps as I saw them.

So, what happened? The Leafs won three of their last four games, but it wasn't enough. Tampa Bay finished two points ahead of them and the Leafs failed to make the playoffs,[11] just as I had predicted. Sometimes even being correct is no fun.

Where Have You Gone, Joe DiMaggio?

Joe DiMaggio played for the New York Yankees between 1936 and 1951. He was a star hitter, leading the Yankees to nine separate World Series championships. He was elected to the Baseball Hall of Fame and is generally regarded as one of the best players ever.

But just one of the best, not the best. For example, his career batting average of 0.3246 ranks 41st-highest of all time among players with at least 3,000 plate appearances.[12] Very impressive, but not right off the charts. Based on that and most other statistics, DiMaggio is one of the greats, but not out of their league.

However, there is one sense in which DiMaggio surpassed them all. In 1941, he had a 56-game hitting streak, meaning he got at least one hit in each of 56 consecutive games. This still stands as the longest hitting streak in the history of Major League Baseball.[13] And not just by a little bit—by a mile. In second place is Willie Keeler, with a 45-game streak spread over parts of the 1896 and 1897 seasons, followed quickly by Pete Rose with 44, Bill Dahlen with 42, George Sisler with 41, and Ty Cobb with 40. (The most recent streak in the top 15 is Chase Utley in 2006, tied for 11th with 35.)

How remarkable is DiMaggio's feat? Over the course of his career, DiMaggio played a total of 1,736 games, in which he had 6,821 at-bats.[14] This works out to an average of nearly four at-bats per game. With a career batting average of 0.3246, his chance of getting at least one hit in any given game works out to about 79 percent. So far, so good. However, this means his chance of getting at least one hit in a specific series of 56 consecutive games is 79 percent multiplied together 56 times, which is just over one chance in half a million.[15] Of course, DiMaggio had *many tries* to achieve his streak, since it could have occurred during any 56 consecutive games over the course of his long career. But the chance that a player with DiMaggio's overall batting average would manage a 56-game hitting streak is less than one chance in a thousand. (In fact, the longest hitting streak we could *expect* him to have would be about 27 games, less than half of 56.)[16]

DiMaggio's streak can't even be equalled in the world of fantasy sports. Since the year 2000, MLB.com has run a Beat the Streak competition in which contestants select one baseball player each day. If that player gets at least one hit that day, the contestant's streak continues. If any contestant's streak lasts beyond DiMaggio's 56 games, they win (appropriately) a cool $5.6 million. However, despite 80 million entries over 17 years of play, the longest streak so far has been 49 games, and the prize remains unclaimed.[17]

What can we make of all of this? Well, we already knew that DiMaggio was a great baseball player. But the numbers show that his 56-game hitting streak went beyond that. He didn't achieve his record based solely on his above-average hitting ability. He also got *really, really lucky*.

On the other hand, it seems even 56 games wasn't enough. Apparently, if he had lasted just one more game, DiMaggio would have earned an extra $10,000[18] for a promotional tie-in with the H.J. Heinz Company, whose slogan was "57 Varieties." *Unluckily*, he didn't quite make it. Ah, what might have been.

Beginner's Luck?

Stephone Anthony had an impressive debut as a rookie linebacker for the New Orleans Saints during the National Football League's 2015 season, with a total of 112 tackles—the most of any Saints rookie in over 30 years. He was selected to that year's Professional Football Writers Association all-rookie team[19] and seemed to have a very promising career ahead of him.

However, his winning ways did not continue. The next season, plagued by injuries and slow development, Anthony only made a disappointing 16 tackles and contributed little to his team. Finally, in September 2017, he was traded to the Miami Dolphins for nothing more than an upcoming fifth-round draft pick. Of course, Anthony might well flourish with his new team and become a star after all—at this writing it is too early to say. But to all appearances it seems that his once-promising career has taken a nosedive.

Such stories are not unusual. In basketball, as of this writing, there are 53 different players who were selected as the NBA Rookie of the Year and have since retired. Twenty-six of them were elected to the Basketball Hall of Fame as acknowledged stars. But the other 27 were not, and had careers of varying success. One extreme case is Woodrow (Woody) Sauldsberry Jr., rookie of the year in 1958 with the Philadelphia Warriors. He hung around the NBA until 1966—with little further success, and a terrible career shooting percentage of just 34.8 percent. The only thing to say here is that young stars sometimes continue on to outstanding careers, and sometimes do not.

This phenomenon has many names. Beginner's luck. Sophomore slump. One-hit wonder. Cup of coffee (referring to someone who was only successful long enough to drink one cup of coffee). It arises not only in sports, but in other fields, like music. For example, guitarist Michael Sembello reached number one in the US and Canada with his 1983 mega-hit "Maniac," but never saw remotely comparable success ever again. This phenomenon is also related to the financial concepts of *regression to the mean* and *mean reversion*, which roughly say that extreme stock prices are often caused by luck and will tend to return to more typical values. Essentially, it all boils down to one observation: if an athlete, team, performer, or stock does particularly well, it could be a marker of actual skill or quality, or it could be just luck—and most likely, it is some combination of the two.[20]

What does this mean in practice? Well, if a student does extremely well on a midterm, then they are probably pretty smart. But they also probably got a bit lucky. So on the final exam, the best guess is that they will do well above average (because they're smart), but not quite as well as on the midterm (because they won't get as lucky). Similarly, if an athlete does particularly well in their first game, they will probably play fairly well throughout the season, but probably not quite as well as that first time. If an investor makes lots of money one year, they will likely make pretty good money the next year, but not quite as good.

Of course, not all examples of beginner's luck need to be explained in such grand terms. Since there are *many people* playing sports at all levels, there are *many tries* for someone to do particularly well their first time out, just by luck alone. And when they do, thanks to *biased observation*, we are more likely to remember their striking success than the struggles of 100 other beginners who all failed. As usual, so much of what we observe—in sports and elsewhere—can be explained by simple luck traps.

CHAPTER 17

Lucky Polls

One area of our modern lives where luck plays an important role is public opinion polls. Pollsters use random samples to study a population's opinions about everything from product preferences to work habits to social attitudes. And most prominently, polls are used to predict the winners and losers of elections. If we are *lucky*, the polls will provide accurate predictions of the final election results. But if we are unlucky, the polls might be significantly off, causing confusion and embarrassment.

Sometimes, polls are spectacularly accurate in predicting election results. For example, just before the 2012 US presidential election, famed analyst Nate Silver sifted through the multitude of election polls and managed to not only correctly predict that Barack Obama would defeat John McCain, but also correctly predict which candidate would win in every single one of the 50 US states[1]—a very impressive triumph indeed.

Other times, polls are less successful. One example is the 2016 "Brexit" referendum about whether or not the United Kingdom should leave the European Union. This referendum was very heavily polled, with most pollsters confident that the Remain side would win[2] by between two and eight percentage points. But when the referendum came, the Leave side won instead, by nearly four percentage points (51.89 percent to 48.11 percent). Many citizens were angry and frustrated that the polls had gotten it so wrong.

Actually, the Brits were already used to bad polling. In their 2015 general election, most polls showed a dead heat between the Conservative and Labour parties. Then the election came, and the Conservatives won by over six points (36.8 percent to 30.4 percent), a very different outcome. This discrepancy was so great that a formal inquiry was called. The subsequent report noted that "On average the final estimates of the polling companies put the Conservatives on 34 percent and Labour on 34 percent. . . . Yet in the event the Conservatives won 38 percent of the vote in Great Britain, Labour 31 percent. . . . In historical terms, the 2015 polls were some of the most inaccurate since election polling first began in the UK in 1945."[3] Others were less diplomatic; the BBC reported, "Following the outcome of the 2015 general election, a mixture of anger and contempt was showered on the pollsters who had spent six weeks suggesting a different result."[4] The *Guardian* called the error "notorious,"[5] and forecasters admitted that "no company consistently showed anything approaching the seven point Conservative lead that happened in reality" and "no one had a good pre-election forecast."[6]

So, which is it? Are polls helpful indications of a nation's mood? Or are they misleading indicators that cause confusion and chaos? How do polls work, anyway?

Random Samples

One challenge for pollsters is basic randomness. If you flip a coin a bunch of times, you will probably not get exactly half heads, but you'll get close to half heads. How close? Well, there are no guarantees. But if you flip a coin some number of times, the percentage of heads will usually—95 percent of the time, or "19 times out of 20"—be within a certain "margin of error" of the true 50 percent probability.

This margin of error has a simple formula: 98 percent divided by the square root of the number of coin flips.

What this means is that if you flip 100 coins, your margin of error is about 10 percent, so you will usually get somewhere between 40 percent and 60 percent heads. With 400 flips, the margin of error is about 5 percent, so you will usually get between 45 percent and 55 percent heads. With 1,000 tosses, the margin of error is about 3 percent, so expect between 47 percent and 53 percent heads. And so on.

And polling companies use this same formula. If a poll samples 800 people and says it is "accurate within 3.5 percentage points, 19 times out of 20," that is because 98 percent divided by the square root of 800 is about 3.5. So, if you flip 800 coins, you will usually get within 3.5 percent of exactly 50 percent heads. By the same token, if you sample 800 people, then your results will usually be within 3.5 percent of their "true" value. Now, this true value isn't necessarily the same as the actual election result, since so many other obstacles could arise: dishonest answers, changing opinions, people not bothering to vote, and so on. But it does quantify the limitations of the poll that are specifically a consequence of its sample size.

These stated margins of error are only correct if you have a good sample. And what makes a sample "good"? It has to be *random*—equally likely to choose any one person as anyone else. In practice, this is very hard to achieve. Some people don't have phones—or have just a cell phone and can't be reached. Others have a phone but refuse to answer it. Or they hang up as soon as they realize a pollster is calling. Or they refuse to answer the pollster's questions when asked.

Indeed, even in the heyday of polling, only about 35 percent or 40 percent of calls actually got a usable response. And nowadays, most poll response rates are less than 10 percent. In short, most people don't want to talk to pollsters. As a result, most poll samples aren't truly random at all.

Now, that is okay, as long as the samples are still *unbiased*. If response rates are low, but are *equally* low for all different types of people, then the sample is still representative of the whole population. In that case, the polling company is very *lucky* and can still predict well. But if some types of people are less likely to respond than other types, this leads to the *biased observation* luck trap and makes polling very difficult indeed. Pollsters try all sorts of tricks to correct for these sampling errors, but none of them provide any guarantees.

The Fiasco of '36

One of the earliest and most dramatic examples of poll failure occurred around the 1936 US presidential election. The incumbent Democratic president, Franklin D. Roosevelt, was facing off against Republican challenger Alf Landon. The influential weekly magazine *Literary Digest* set out to predict who would win the election.

In an incredibly large-scale effort, the magazine mailed survey questions to ten million different households and received 2.4 million replies. (To put this in context, the election itself recorded a total of just 44.4 million votes, so they probably received replies from more than one voter in 20.) After painstakingly tabulating all of this information, the magazine announced in its issue of October 31, 1936, that Landon would win handily, with 55 percent of the popular vote to Roosevelt's 41 percent, and with 370 of the 531 electoral college votes.

Boy, were they wrong. On election day Roosevelt won 60.8 percent of the popular vote to Landon's 36.5 percent. And Roosevelt won 523 of the 531 electoral college votes—the most lopsided result in US history.[7] The *Literary Digest* prediction was so far off, and so absurd in retrospect, that it lost all credibility with its readers. A year and a half later the magazine folded and was never heard from again.[8]

By contrast, George Gallup conducted a much smaller survey of just 50,000 voters. Based on this, he predicted that Roosevelt would win 56 percent of the popular vote to Landon's 44 percent,[9] which was quite close to the actual result. Gallup went on to become one of the most famous and successful public pollsters of all time.

How could Gallup predict so well with a sample of 50,000, while the *Literary Digest* did so poorly with a sample of 2.4 million? Because Gallup's sample, though smaller, was better. He was careful to make his sample as representative of the population as possible, surveying voters rich and poor and far and wide. The magazine's poll, on the other hand, had surveyed only its own readers, along with registered automobile owners and telephone users—all fairly "elite" groups back in 1936 and, it seems, more likely to favour Republicans. Furthermore, just 24 percent of these people bothered to mail back the survey, and those who did were probably more affluent or angry or otherwise pro-Republican than the 76 percent who didn't bother.[10] The two polls stand as a textbook example of how an *unbiased* sample is more important than a large one.

Trumped

The most dramatic poll failure in recent years has got to be the one that accompanied the 2016 US presidential election. In the days leading up to the vote, most polls showed Hillary Clinton ahead of Donald Trump by about four percentage points.[11] In fact, Clinton did win the popular vote, though only by about 2.1 percentage points.[12] The vagaries of the electoral college's state-by-state system allowed Trump to squeak through as the next president of the United States.

In terms of the popular vote, then, the pre-election polls were off by approximately 1.9 points. Simply the margin of error, right? Wrong. So many polls were taken in the days before the election that their combined sample size was well over 30,000,[13] leading to a combined margin of error of about one-half of one percent—much less than their actual 2.1 percent error.

Why was the error so large? Simple: *biased observation*. The situation was nicely summarized by CNN commentator Van Jones. In an interview he gave while visiting Canada, Jones noted that Trump's election "was a repudiation of a lot of stuff in our country." What stuff? In addition to such standards as political insiders, banking elites, foreigners, academics, and Hollywood stars, Jones also listed "this overhang of pollsters that think they know everything."[14]

That line struck a chord with me. If Trump voters resented pollsters (among others), they might be even less likely to respond to polls than Clinton voters. Could that make a difference to the results?

I quickly began calculating. I imagined a scenario in which a polling company phoned a perfectly representative sample of voters and all respondents were completely truthful. The only problem? Not everyone agreed to respond. If the response rate was 10 percent for Clinton and other non-Trump supporters, then how much lower would it have to be for the Trump supporters for the pollster to conclude that Clinton was ahead by four percentage points? The answer was 0.4 percent lower. If the response rate was 10 percent for non-Trump supporters, and 9.6 percent for Trump supporters, then that alone would explain all of the pollsters' errors.[15] That's the power of *biased observation* at work.

Sweet Home Alabama

While I was finishing this chapter, a minor-seeming election in the United States took on great significance. The state of Alabama held a special vote to replace one of its senators, Jeff Sessions, who had resigned to become US Attorney General.

Alabama is one of the most Republican-leaning of US states— Trump took 62 percent of the vote in the 2016 presidential race[16]— so it was considered certain that a Republican would win the vote. However, the Republican candidate ended up being Roy Moore, a controversial far-right judge with extreme opinions and limited appeal. And then, during the campaign, nine different women accused Moore of sexual assault or sexual misconduct. These difficulties changed the race from a landslide to a nail-biter. And since the senate was almost equally divided between Republicans and Democrats, the vote took on major national importance, with President Trump publicly endorsing Moore.[17]

So, what did the polls say? In short, everything! On the eve of the election, one poll showed Moore up by nine percentage points, one found him trailing by ten, and still another suggested an exact tie[18]—a huge spread. Most of the other polls showed Moore slightly ahead, but the variation was a genuine source of uncertainty. Despite the national importance, the multiple polls, and the extreme attention being paid to the election, the reality was that no one knew what was going on. Even Nate Silver himself could do no more than write an article called "What the Hell Is Happening with These Alabama Polls?"[19] As one frustrated observer aptly tweeted (in words that called to mind an earlier Trump tweet on his proposed immigration ban), "I am calling for a complete and total shutdown of polling until we can figure out what the hell is going on."[20]

When the votes were finally counted on December 12, 2017, Moore took an early lead. But this was a form of *biased observation*: it takes longer to count votes in metropolitan areas, where Moore was weakest. Finally, in the late evening, the result became clear: Moore had lost. The new Alabama senator would be not Moore, but his Democratic opponent, lawyer Doug Jones, who won by a margin of about 1.5 points (49.9 percent to 48.4 percent). For the first time in 25 years, Alabama had elected a Democratic senator. The polls may have been confusing, but the final result was anything but.

Polling Reflections

When polls do not predict well, they are heavily criticized. I recently went to lunch with some professional statisticians before a research seminar. Talk turned to the polls, and how they had failed to predict Trump's victory. One statistician felt such dismay that he could not conceive how that could have happened. He said he wanted to talk to those pollsters "after a few drinks," at which point they would hopefully reveal to him just how they had all failed so badly. The implication was that the pollsters had been so incompetent that there must be some big secret required to explain it.

Was this statistician being fair to the pollsters? I don't think so. With such low response rates, the difficulties of obtaining truly random samples, and respondents' reluctance to answer truthfully, conducting accurate polls to predict close elections is actually a very challenging task. Anyone can predict a landslide, but when the result is within a few percentage points, all bets are off—no matter how clever and brilliant the pollster.

After hearing my various comments about the limitations of polls, a radio interviewer once asked me if I liked polls, and seemed genuinely surprised when I replied in the affirmative.[21] What about all those polling problems and errors and inaccuracies? he wanted to know. Don't they make you want to hate the polls, and the pollsters too? I replied sanguinely that despite all the challenges, polls remain virtually the only method available to reasonably gauge the preferences or intentions of an entire population. Without polls, we would merely be *guessing* what the country thinks. Or, worse, we would rely on the opinions of a few friends, leading to *biased observation* and wildly inaccurate results. Polls are far from perfect, but at least they are providing *some* reasonable estimates, which more often than not are at least somewhat close to the truth. This extra information is then used by candidates to adjust their campaigning, citizens to strategize their votes, politicians to follow majority opinions, businesses to anticipate changes, and more. And that is all a lot more valuable than some people realize.

But despite their extreme utility and importance, polls are also very challenging—and sometimes completely wrong.

Lucky Sayings

We are all familiar with common sayings. More accurately called aphorisms, they are defined as "a terse saying embodying a general truth."[1] Some of these sayings may be silly, or even meaningless. But many of them do seem to contain wise words that offer useful insights into the workings of the world.

A number of these sayings are somehow related to luck. One way or another, they make assertions or offer insights into the influence of luck on our daily lives. Now that we have a better understanding of the nature of randomness and chance, can we use that to better understand, and more accurately evaluate, these various sayings? Sometimes.

Better safe than sorry. This is a common refrain. Decoded, it says we should always take extra precautions rather than take risks. I say, not necessarily! If a bad outcome is sufficiently unlikely, it might be better not to worry about it at all. For example, we have already seen that only about one airline flight in five million has a fatal accident. Similarly, fewer than one child in half a million will be abducted this year by a stranger with evil intentions.[2] Should we live in fear and avoid flying in the rain and prevent our children from walking to school, all to avoid these extremely unlikely eventualities? I think not. Sometimes excessive stress and worry just aren't merited by such unlikely possibilities. **Rating: False.**

Fortune favours the bold. This is an old Latin proverb that has also been variously quoted as fortune favouring the "brave" or the "strong."[3] (Or, as *Star Trek*'s always-modest Captain Kirk once put it, "May fortune favour the foolish.")[4] Now, I don't actually believe that fortune favours anyone at all—I think it is just random. On the other hand, in some circumstances, perhaps including sports and battles, being bold or brave or strong (though—sorry, Kirk—not foolish) may sometimes help to provide you with a greater chance of success. Not always, of course—it is also possible to be too bold and take too many risks and then suffer greatly. Different circumstances can lead to different conclusions, with no clear rule. **Rating: It depends.**

Tomorrow is another day. This is one of those sayings that, on the surface, is completely banal—of course tomorrow is another day. (Well, unless the world ends tonight, which I think and hope isn't very likely.) But it also has a deeper meaning. It is saying that even if bad things happened today, you might have better luck tomorrow and things might improve. Is it really so? Well, if today's troubles were because of your own specific life situation, then they likely will not change tomorrow. But so many of our problems are caused by external random events—that is, by bad luck—and for those problems, the bad luck of today might well be replaced by good luck tomorrow. **Rating: Mostly true.**

Good things come to those who wait. This saying has a philosophical side, encouraging patience and serenity even when goals seem out of reach. But it also has a practical meaning, that if you keep on trying or waiting, you will eventually get what you hoped for. And why might that be? Well, because of *many tries*. If you wait a long time, then finally one day you will probably find success. (Indeed, the winner of the 2016 Nobel Prize in Literature himself once wrote about the importance of waiting for a simple twist of fate.[5]) **Rating: Often true.**

The luck of the Irish. This saying is often accompanied by tales of lucky charms and leprechauns. It implies that people from Ireland inherently receive more good luck than the rest of us. Now, this seems implausible at face value. Surely the universe's fortunes, however they are controlled and decided, do not actively favour one nationality over any other. And to be frank, the history of Ireland is not entirely a lucky one—most obviously, the potato famine of the late 1840s killed about a million Irish and forced about as many to emigrate to other lands. Some have claimed that this saying really refers to a positive Irish attitude,[6] or perhaps to certain specific Irish successes during the 19th-century US gold rush.[7] But it is usually taken to imply that Irish people are magically more fortunate than the rest of us, which seems absurd and is not supported by any evidence. **Rating: False.**

Nobody said that life is fair. This is one of those sayings that can be very annoying. It is usually uttered by some smartass right after you suffer a bad break. I do *not* recommend using this saying at such times! On the other hand, even people who believe in some sort of fate, karma, or all-powerful god will still surely admit that lots of unfair *bad luck* is experienced by lots of good people who deserve better. As much as I wish that life's luck traffic were directed with a careful eye towards making everything fair and balanced, it just isn't so. If something bad happens to you, you might actually deserve it, but you might not. And if your annoying neighbour wins a lottery jackpot, he might not deserve that either. Luck certainly isn't always fair. **Rating: True.**

Make your own luck. This saying implies that luck is (often) not a magical power that comes from without. Rather, what appears to be lucky is often the result of hard work, careful planning, appropriate caution, and painstakingly developed skills. For example, Richard Wiseman spent ten years monitoring and interviewing subjects for his book *The Luck Factor*, and he concluded, "The findings have revealed that luck is not a magical ability or the result of random chance. Nor are people born lucky or unlucky. Instead, although lucky and unlucky people have almost no insight into the real causes of their good and bad luck, their thoughts and behavior are responsible for much of their fortune."[8] In short, people often create their own good and bad luck through their various actions. **Rating: Often true.**

It is better to be born lucky than rich. Related sayings go back as far as the 1639 quotation "Better to have good fortune than be a rich man's child."[9] And 1930s baseball pitching great Lefty Gomez was fond of saying, "I'd rather be lucky than good."[10] These sayings are usually taken to mean that luck is more important than any money or skill can ever be. And really, who can argue with that? If you're rich, you could have a stroke of bad luck and lose everything. And even if you're good at something like baseball, on any given day a ball could bounce the wrong way, your timing could be off by a fraction of a second, or the other team could manage to snatch victory from the jaws of defeat. But if you're *lucky*, well, things will work out well for you and you will accumulate wealth and victories just by luck alone. In fact, in the 1990s the sports analytics company TruMedia Networks invented a "luck statistic" that tried to measure how lucky a team had been, in the sense of how many games it actually won compared to how many games it "should" have won given the differences between points scored and allowed.[11] In sports, as in everything else, luck matters.

A different perspective comes from Lamar Gillett, a World War

II fighter pilot who achieved the rare success of shooting down a Japanese Zero aircraft. He later told the journalist Thomas McKelvey Cleaver, "I was lucky I was behind the Zero instead of in front of him. I was lucky when I landed back at Clark that the guys who were shooting at me didn't give enough lead. I was lucky my C.O. on Bataan sent me to Corregidor to get the chewing-out he was in for and wasn't there for the shelling that killed him."[12] Gillett understood the importance of luck in battle. His conclusion? "It's better to be lucky than good."

So, if the saying were "Luck is more important than money or skill," I would have to rate it as **True**. On the other hand, as we have just seen, there is no reason to think that people or teams are *inherently* lucky. Even if you have been lucky for a while, it may not last, and bad luck might be just around the corner. No one is "born lucky"; people just experience lucky or unlucky events as life goes on. Yes, luck is more important than money or skill, but no, people are not *born* with luck. **Rating: Partly true.**

The older you get, the more everybody reminds you of someone else. I can't remember where I first heard this, but it certainly rings true for me. When I was younger, every new person I met seemed entirely original, different from every other person. But in recent years, when I meet someone, I frequently notice their similarities to people I've known before. Either they look similar, or they speak in the same style, or they have similar attitudes—for better or for worse. So, is it true? Do people seem more familiar just because we're older? Yes indeed. And there is a very simple explanation for it—it is the *many tries* effect. When you are young, you don't know too many people too well. It would be incredible luck if a new friend seemed similar to one of the few people you were already familiar with. But when you're older, you have spent many years meeting many people. So there are lots of people in your mind that your new friend could resemble. It is much less surprising that they are similar to *someone*. So the next time an older person tells you that you remind them of someone else, tell them that is great, but *out of how many* other people that they know? **Rating: True.**

Bad things come in threes. This is a classic case of *biased observation*. If one bad thing happens to you, you will suffer through it without looking for a pattern. If a second bad thing happens, you will suffer through that too. But if you encounter a *third* negative event, you will be struck by the coincidence and will tell everyone about your string of bad luck. In this way, everyone remembers when three bad things happen in a row, but no one notices when there are only one or two. And there is no force that makes a third negative event more likely just because you have recently had two previous unrelated moments of misery. **Rating: False.**

When it rains, it pours. This is the flip side of the previous saying. If one or two bad things happen to you, you won't notice or comment upon the pattern. But if *lots* of bad things happen, you will be overwhelmed by the onslaught and remember it for a long time. More *biased observation*. Ironically, this saying isn't even literally true—often a light rain dissipates without producing any downpour. **Rating: False.**

A stitch in time saves nine. This old classic is literally about keeping clothes in good repair to avoid later damage. But more broadly, it means that if you prepare well and plan carefully in advance, you will have fewer problems and more success in the future. No one can argue with that. And what is the connection to luck? Well, if someone *appears* to be lucky, it could just be random luck, but it could instead demonstrate that they were better prepared and thus had a greater chance of success. **Rating: True.**

Out of sight, out of mind. We are certainly more likely to forget things once they are gone. (Yes, even despite the competing saying that "Absence makes the heart grow fonder.") How is this connected to luck? Well, it helps to explain the luck trap of *biased observation*. When we observe some special or dramatic outcome, we often forget all of the other times when that outcome did not occur. So its occurrence might be just dumb luck, without any special meaning. To evaluate any outcome properly, always wonder *out of how many* different times it actually occurred. **Rating: True.**

Don't quit your day job. This is usually used as a cruel or amusing (depending on your perspective) jab at some kind of performer to imply that they will never make much money with their performances so they had better continue to earn a living more conventionally. (I have even used this phrase in my own improvisational comedy player profile, to occasional amusement.)[13] More broadly it implies that cultural performing such as music or comedy is usually not very lucrative. Is it true? Well, we all hear stories of musicians ranging from B.B. King and Ringo Starr to Jewel and Shania Twain whose musical success transported them from poverty to mega-millionaire status.[14] This gives many aspiring performers optimism that they will soon have similar success. But this optimism arises from the luck trap of *biased observation*. We only hear about the small number of musicians who get very rich, while hearing nothing about the tens of millions who try and fail. The computer revolution has only amplified this effect: as the *New York Times* recently put it, "Extraordinary lives look like the norm on the internet."[15] Once this luck trap is stripped away, it unfortunately follows that most aspiring performers will not make much money, so for nearly all of us, a regular, conventional job is indeed our best chance for monetary success. **Rating: Mostly true (unfortunately).**

A little suffering is good for the soul. This saying means that misfortune can help to build character and teach valuable life lessons—sort of like Justice John Roberts's commencement speech wishing the students bad luck. That sounds plausible, though it's hard to verify precisely. On the other hand, this saying is also sometimes taken to mean that if you suffer today, then "karma" will guarantee you better luck tomorrow, balancing everything out in the end. Unfortunately, there is no evidence of any such balancing force: your suffering today can't actually protect you from later troubles. Therefore, this saying can be true or false depending on the interpretation. **Rating: It depends.**

With proper medication you can cure a cold in seven days, but left to its own devices, it takes a week. I'd heard that the great mathematician and philosopher Bertrand Russell once said this, though I can't find a reference. Anyway, its point is that colds tend to go away in about a week (give or take) all by themselves. So any medicine that "cures" your cold in that time isn't really doing anything at all. Put another way, curing a cold in seven days is an *extra-large target* whose accomplishment means nothing. The body's natural healing powers provide an *alternate cause* (besides the medicine) for your return to health. This saying illustrates an important general principle. Whenever someone claims that they can achieve a certain result or accomplishment, reduce a certain problem, or make something happen very often or very quickly, you should always ask: Compared to what? How does your achievement or outcome compare to what *would* have happened without intervention, or in an average person, or with the previous techniques, or with some competing intervention? You should only be impressed if their course of action shows a true improvement over what would already have been achieved by other means. **Rating: True.**

Every cloud has a silver lining. This saying is primarily used to cheer people up after something bad happens. It says that even bad things are always accompanied by something positive. If we interpret that statement precisely, then it must be true—though just barely—since every situation has *some* good aspects (but in some cases, the good aspects are very minimal indeed). The connection to luck is that most situations have both positive and negative aspects, so they can be interpreted in different ways, leading us to draw different conclusions, which is the essence of the *different meaning* luck trap. However, some situations are quite awful, with only the slimmest of positive aspects to uncover. **Rating: True (barely).**

Everything happens for a reason. This saying can provide great comfort when dealing with misfortune. It is a modified version of the silver lining. It tries to assure us that even if things appear bleak, there is some sort of underlying logic involved, some reasonable justification for our troubles, and implicitly some unseen benefit to the unpleasant outcome. I wish it were so. However, I don't see any clear evidence at all for this hopeful perspective. Our lives are subject to a multitude of random influences. And unfortunately some of these influences are just bits of dumb luck leading to negative consequences, with no balancing force to make it all right again and no reason or explanation to comfort us either. **Rating: False.**

Truth is stranger than fiction. This is often said when we're marvelling at some very unusual outcome, and pondering how strange our world can be. But it also contains a deeper meaning, about fiction itself. Of course, fiction can involve all sorts of different imaginative aspects, from space aliens to talking animals to fantasy worlds to magic powers, seemingly without limit. However, as we have seen, fiction has its own constraints. It has to strike readers as somehow plausible or logical or reasonable, or they will dismiss it. It has to have some sort of meaning or structure or narrative arc, in which the pieces come together in certain ways. And most importantly, our search for meaning and magic in the luck of fiction actually serves to limit fiction's possibilities in some ways. In most fiction, the heroes have to triumph, suffering needs a purpose, everything happens for a reason, and the good guys eventually win. Truth is also constrained—by the laws of science that govern our universe. But truth does not have to follow fiction's rules. In real life, sometimes bad things happen for no reason, innocent people suffer unnecessarily, evil conquers all, the riches go to the person who least deserves them, justice is not served, and none of it makes any narrative sense at all. **Rating: Sometimes true.**

There are no passengers on Spaceship Earth. We are all crew. This observation was uttered by Marshall McLuhan in response to Buckminster Fuller's 1963 book *Operating Manual for Spaceship Earth*,[16] and it has always stuck with me. Indeed, what is planet Earth but a large vessel hurtling through space, with limited resources and an uncertain future? Any one of us might affect that future through our various actions. None of us has the luxury of just sitting back and watching some other "crewmembers" at work. This saying illustrates that our actions have consequences. What may seem to be just dumb luck is sometimes actually the result of an *alternate cause*, based on specific actions by specific people. In other words, as we have already seen, in one way or another, we often have to take actions in order to make our own luck. **Rating: True.**

It seems there is a luck-related saying for nearly every occasion and situation. Some of them are helpful, and some of them are funny, and some of them are absurd. None of them should ever be taken too literally. However, they do sometimes influence our perspective, how we think about things, how we feel. They can comfort us, enlighten us, calm us, or annoy us. And with a better understanding of luck and chance, sometimes we can see them in a new light. At the very least, we can spot which ones are true and which ones are, well, a bunch of hooey.

CHAPTER 19

Justice Luck

Of all the areas in which luck needs to be assessed, perhaps none is more important than crime. Will a perpetrator get caught, or be lucky enough to get away with it? Is the evidence against a defendant sufficient to prove their guilt "beyond a reasonable doubt"? Or is the evidence misleading, arising just because of, well, really bad luck? Issues of capture, conviction, sentencing, prison, of justice itself—indeed, a defendant's entire future—can hinge on twists of luck. Sometimes, these questions involve subtle issues of probability and inference. Other times, they can even be amusing.

Bumbling Bandits

On August 29, 2017, at about 5:30 p.m., two men entered an Irish pub in Baltimore.[1] But these were no ordinary men. Their faces were hidden by masks. They pointed guns at the workers behind the counter and forced them to hand over the money in the cash register. At that point, they hurried off, with the proceeds of another robbery safely in hand.

There was just one problem. By *incredibly bad luck*, the large, friendly party in the pub's main room was in honour of a police sergeant who was retiring after many years on the force. The sergeant's many friends and colleagues were all in attendance. In short, the place was crawling with cops. Upon hearing of the robbery, several officers chased the retreating robbers, and quickly caught and arrested them. Instead of making away with the cash, they got charged with armed robbery and theft.

Call it a robbery gone bad, for one reason only: luck.

Those robbers have nothing on two Alberta women who were serving two-year sentences for theft, after a history of violence and weapons offences. At 8:40 p.m. on October 2, 2017, they scaled the east fence of the Edmonton Institution for Women and escaped—the first such escape in nearly three years.[2] A police search of the immediate area turned up nothing, and it seemed the women had gotten away. What would become of them?

Unluckily for them, their escape generated so much publicity that someone recognized them the next day and notified police, who quickly made an arrest. In total, they were at large for just under 24 hours—not the most successful prison escape in history.

And most dramatically of all, where were they arrested? In an escape room! That's right, in one of those interactive team-based fantasy games, the kind where a group of people work together to try to solve puzzles and search for clues in an effort to escape from the room before their time runs out.

It seems that the prisoners, excited and pleased after their big escape from an actual prison, decided it would be fitting to try their luck at a rather different kind of escape. But *as luck would have it*, one of the other game players had a keen eye not just for escape games, but for actual prison escapees too. The game owner wisely remarked, "It's all kinds of ironic."

Do these stories tell us anything new about luck? Not really. In each case, nothing more than a *many tries* luck trap is at work. Out of all the robbery attempts at pubs around the world, it is not surprising that one of them came upon a police party. And out of all the times that prison escapees were recaptured, it's no great shock that one of them involved an escape-room game. There is no deep meaning, or fundamental principle, involved in such stories.

But they're still rather amusing.

Moustaches and Ponytails

Considerations of luck and probability sometimes play an essential role in actual criminal convictions. A classic example arose on June 18, 1964, in Los Angeles. An elderly lady was pushed down in an alley and her purse was stolen. Witnesses said a young Caucasian woman with a dark blond ponytail ran away with the purse and got into a yellow car driven by a black man who had a beard and moustache. Four days later, Malcolm and Janet Collins were arrested, primarily because they fit these same characteristics (at least mostly—Janet's hair was apparently light blond rather than dark blond, but that difference was ignored).

At trial, the prosecutor called "a mathematics instructor at a nearby state college" (whose identity I have been unable to locate—it seems that everyone forgets the poor mathematician). The prosecutor told the mathematics instructor to assume certain "conservative" probabilities:

- Black man with a beard: 1 out of 10
- Man with moustache: 1 out of 4
- White woman with blond hair: 1 out of 3
- Woman with a ponytail: 1 out of 10
- Interracial couple in a car: 1 out of 1,000
- Yellow car: 1 out of 10

The mathematics instructor then computed the probability that a random couple would satisfy all of these criteria, by multiplying the various factors together:

$$(1/10) \times (1/4) \times (1/3) \times (1/10) \times (1/1000) \times (1/10) = 1/12,000,000$$

He thus asserted that there was just one chance in 12 million that a couple would have these same characteristics by luck alone. Malcolm Collins was convicted at trial, primarily based on this probability.

Was this sufficient evidence for conviction? Or was it just Collins's bad luck?

Several problems with the calculation immediately present themselves. For one thing, the stated probabilities were just asserted, without any evidence to back them up, and they might well have been faulty (and not necessarily "conservative"). But more importantly, many of them *go together*. Most men with beards also have moustaches. A black man and a white woman necessarily form an interracial couple. Women with blond hair might perhaps be more likely to have ponytails. And so on.

Instead, a more reasonable interpretation of the assumed probabilities is that one car out of 1,000 contains an interracial couple, and of those, one in ten includes a man with a beard and moustache, and one in ten includes a woman with a blond ponytail, and furthermore one in ten such cars is yellow. If so, then the probability of a randomly chosen car fitting the witnesses' description would be:

$$(1/1,000) \times (1/10) \times (1/10) \times (1/10) = 1/1,000,000$$

or one chance in just one million, not 12 million. And even this calculation is rather suspect, relying as it does on lots of unjustified assumptions.

But most importantly, there is a *shotgun effect* at work here. Namely, out of all the cars that could have been spotted, one of them happened to fit the witnesses' description. Does that prove an effect (in this case, that Collins was guilty), or could it be just luck?

Well, in 1964, Los Angeles County had a population of about

6.5 million people. This probably corresponds to over a million different households, each with a car that could have matched the description. So, with over a million cars to choose from (to say nothing of cars from other nearby counties that could also have driven in), it would not be surprising if *one* of them matched these one-in-a-million odds just by chance. That is, it could well be nothing more than dumb luck, proving nothing.

This suggests that Collins should not have been convicted on this basis. As it happened, on appeal four years later, the Supreme Court of California agreed. Their decision began, "We deal here with the novel question whether evidence of mathematical probability has been properly introduced and used," and ended, "We conclude that on the record before us defendant should not have had his guilt determined by the odds and that he is entitled to a new trial. We reverse the judgment."[3]

Does this mean that Collins was necessarily innocent? No. The probabilities presented were not sufficient for conviction, but they don't prove innocence either. In fact, there may be some other evidence that Collins actually did commit the robbery. For one thing, his wife, Janet, was charged alongside Malcolm and did not appeal her conviction. For another, they paid off $35 worth of traffic fines the day after the robbery, and the victim claimed that there was $35 to $40 in her purse when it was stolen. This was suspicious, since there was evidence that the Collinses only had $12 between them when they were married two weeks earlier and hadn't earned too much money since then. Most dramatically, when questioned about the origin of that money, Malcolm Collins said he won it gambling, while Janet Collins said it came from her salary. Very suspicious!

So perhaps the Collinses were guilty after all. But not because of the probability calculation and matching the witnesses' description—that part could have been just bad luck.

The Sad Tale of Sally Clark

Nothing illustrates the challenges of luck and justice better than the remarkable case of Sally Clark, an English solicitor.[4] After a successful legal career, she gave birth to a son. Three months later, the son died of apparent asphyxiation. The cause was unclear, so it was declared to be sudden infant death syndrome, or SIDS (also called "cot death" or "crib death"), the tragic situation in which an infant suddenly stops breathing (usually while sleeping at night) and dies, for no apparent reason.

That would be sad enough, but the story gets worse. Clark had a second son 17 months later who also died of apparent asphyxiation, also of unknown cause, at the age of eight weeks. Clark had thus, tragically, lost two babies in less than two years.

The two deaths raised the suspicion of authorities, who were then faced with a dilemma. Was Clark simply the victim of very, very *bad luck*? Or, was there some *other* cause of her sons' deaths? In particular, had she perhaps *murdered* them both by suffocating them in their sleep?

The prosecutors decided that she had, and Clark was charged with double infanticide. At her trial, a pediatrician testified that the probability of two children from an affluent family suffering cot death was a staggering one chance in 73 million, since a child has one chance in 8,543 of dying from SIDS, and 8,543 × 8,543 is about 73 million. Thus, the prosecution argued, the two deaths surely weren't due to luck. Rather, Clark was a murderer.

The jury agreed. Clark was convicted and sent to prison. Her third son was taken away from her. She was reviled in the press and harassed by other prisoners. She became an alcoholic, which eventually killed her. And still, the question remained: Had the jury gotten it right? Had Clark really killed her two sons? Or was it just luck?

Were there any luck traps at work here? Yes indeed.

First of all, there was some *false reporting*. The fraction of infants who get SIDS is actually about one in 1,303. The pediatrician had adjusted this to one in 8,543 by taking into account certain factors that lower the probability (the family had no smokers, at least one person in the household with a job, and a mother over 26 years old), but ignoring certain other factors that raise the probability (notably, that SIDS is about twice as common in boys as in girls).

Second of all, there were some *facts that go together*. Specifically, it is known that SIDS tends to run in families, and indeed it has been estimated that a family with one SIDS death is between five and ten times more likely to have a second one.[5] So the probability of two SIDS deaths is *not* the same as multiplying the two probabilities together. Rather, we should multiply that probability by about five or ten.

After fixing these two mistakes, the probability of two SIDS deaths by luck alone switches from one in 73 million to about one chance in 1,303 × 1,303/10, or about one in 170,000. Now, that is still pretty darn unlikely, but much less so than before. And the biggest luck trap is yet to come.

Namely, there is a *shotgun effect* here. The Clark family had two SIDS deaths, but this is just one family out of millions of families in England. The fact that *one* such family had two SIDS deaths is tragic but not so surprising, and certainly no basis for a murder conviction.

Upon hearing about the conviction, the venerable Royal Statistical Society noted that the conclusion of guilt was "statistically invalid" and "the case of *R v. Sally Clark* is one example of a medical expert witness making a serious statistical error, one which may have had a profound effect on the outcome of the case."[6] Eventually, justice was done. The pediatrician was ruled to be guilty of "serious professional misconduct," and his evidence "misleading and incorrect." Clark was acquitted and released, but only on her second appeal after spending more than three years in jail. The perils of misleading statistics, indeed.

Five Times Too Many

In the Sally Clark case, having two babies who both died of SIDS was actually not that suspicious after all. But what if it had been five babies instead? Would that have been cause for alarm?

Such is the story of Waneta and Tim Hoyt. They had five babies in New York State between 1965 and 1971. Incredibly, all five of them died in infancy (at 3, 28, 1.5, 2.5, and 2.5 months old, respectively). The deaths were all identified as SIDS, and indeed a pediatrician used them to publish a scholarly article about SIDS' strong genetic link-age—that is, the large extent to which SIDS tends to run in families.[7] Apparently, no foul play was suspected, and in fact in 1977 the Hoyts were allowed to adopt a son (who survived to adulthood).

Years later, in the 1980s, some prosecutors and pathologists became suspicious and investigated. Eventually, Waneta Hoyt tear-fully confessed to suffocating all five of the children to stop them from crying. She later recanted her confession, but that was too late. She was convicted in 1995 of five murders[8] and died in prison in 1998 at the age of 52.

Should her murderous ways have been detected sooner? Probably. We have already seen that there is some genetic linkage between SIDS deaths, and indeed having one SIDS death in the family makes a second SIDS death five to ten times more likely. Nevertheless, five SIDS deaths in one family is still very suspicious. Using the above logic, the chance of that happening by luck alone is something like one chance in

$$1,303 \times (1,303/10) \times (1,303/10) \times (1,303/10) \times (1,303/10)$$

which is about one chance in 375 billion. This is sufficiently unlikely to overcome all possible luck traps. It should have raised alarm bells at the time, and certainly should have been investigated much sooner than it was. The fact that it took 20 years for authorities to notice the problem was *very bad luck* indeed.

Killer Nursing

Lucia de Berk was a nurse who worked on three different hospital wards in The Hague, Netherlands, and was generally well liked and respected. That is, until one day when she was arrested and charged with multiple murders. It was alleged that on at least ten occasions, she had poisoned her patients in an attempt to end their lives.

Our first reaction might be that such a crime is impossible. Nurses are wonderful people who make great sacrifices and endure much hardship and unpleasantness, all in an effort to make their patients more comfortable. Surely no nurse would ever try to harm a patient? But while I was revising this chapter, a news story broke about a nurse, Elizabeth Wettlaufer, who confessed to killing eight elderly patients over a period of nine years, within 150 kilometres of my house.[9] She said that while caring for the patients, she felt a "red surge" that she figured "must be God" telling her that the patients had to die, so she gave them fatal doses of insulin. Her actions went completely undetected for years, until she chose to enrol herself in a drug rehabilitation program. At that point, her strange comments about the deaths caused the hospital staff to contact police, who investigated and eventually extracted her confession. So it seems that nurses sometimes do harm patients after all. The question is: Did Lucia de Berk commit similar crimes, or didn't she?

The evidence against de Berk was primarily about probabilities. A statistician for the prosecution had studied various "incidents" (patient deaths or near-deaths) in three hospital wards where she worked. He reported that she had been on duty for 14 of the 27 such incidents (51.9 percent), despite working just 201 of the 1,704 nursing shifts in those wards (11.8 percent). He asserted that there was just one chance in 342 million that such an imbalance would occur by chance alone.[10] This seemed to demonstrate that the string of misfortune during de Berk's hospital shifts couldn't be just luck. The only other option was that de Berk had deliberately poisoned these patients—perhaps as so-called "mercy killings." Although she protested her innocence, de Berk was ultimately convicted of seven murders, plus three attempted murders.

Was there any other evidence against de Berk? Well, a little. Two patients were found to have elevated levels of medicines that may have harmed them and might perhaps have been administered by de Berk.[11] Also, it was held against de Berk that she had previous worked as a prostitute[12] and had supposedly stolen two books from a library.[13] And most damning of all, she had written some suspicious notes in her diary, including that she "will take this secret with me into the grave" and "today I gave in to my compulsions" (which she later claimed referred only to tarot card reading). But this evidence was circumstantial and unconvincing. The main evidence against de Berk was the one-in-342-million probability.

On that basis, was her conviction justified? Or were some luck traps involved?

We might start by asking whether the incident counts were accurate. There was some debate about whether all of these incidents had actually taken place *during* de Berk's shifts, as opposed to just before or just after. Perhaps some *false reporting* was involved.

And, was there an *extra-large target*? Could be. It seems that the definition of "near-death" was adjusted post hoc to include more incidents during de Berk's shifts. Indeed, in some cases, deaths that had previously been classified as "natural" were suddenly reclassified as "unnatural" when it was realized that de Berk had been on duty for them.

Was there any *hidden help*? Possibly. It seems that de Berk may have been assigned to more elderly and terminally ill patients than most nurses, due to her particular expertise and interests. If so, this would provide something of an *alternative explanation* for a high number of incidents.

However, as is often the case, the single most important luck trap is the *shotgun effect*. All we know is that one nurse, somewhere, had experienced an extra-large number of patient incidents. But how many different nurses *could* have had this same experience? To be fair, the prosecution statistician had tried to account for this issue. What he did was multiply his probabilities by 27, which was the number of nurses in one of the hospitals where de Berk worked. That is a start, but I would argue that he really should have multiplied not just by 27 but by the total number of nurses in the Netherlands—or perhaps even in the entire world.

Gradually, after de Berk's initial conviction, various statisticians started to voice these objections.[14] Enough doubts were raised about the probability calculations that her conviction was upheld on appeal mostly on *other* grounds, notably the elevated medicine levels. Even this evidence was later disproven, and the convictions were overturned on a second appeal. Lucia de Berk received an apology and is now a free woman.[15]

So, did she do it? Did nurse Lucia de Berk secretly kill her patients, perhaps to relieve their suffering in their final days? We cannot be sure. But the evidence against her was insufficient for conviction. The probabilities sounded convincing, but they fell apart in the face of the sharpshooter's luck traps.

CHAPTER 20

Astrological Luck

As part of an effort to control and explain life's randomness and luck, many people turn to astrology. It seems that virtually every newspaper publishes some sort of horoscope. According to one survey, 21 percent of adults reported they read their horoscopes often or fairly often, and 25 percent thought of astrology as "very scientific." (To be fair, only 7 percent think the same of horoscopes, so perhaps some of them are confusing astrology with the actual science of astronomy.)[1] And believers come from all corners—I have known high-level professors, researchers, and scientists who still feel their lives are influenced by astrological forces.

The Chinese have their own popular astrological system based on 12-year cycles.[2] As one observer put it, "If you ask people in China if they believe in the zodiac, many will initially say, 'No, no. We are modern.' But if you ask them when they want to have children, they'll say, 'Hey, it's not a bad idea to have a Dragon baby.'"[3] In recent years, the Chinese have also started to embrace Western astrology—one recent advertisement for a job as a sales representative in Changzhou, China, ended by saying, "Scorpios, Capricorns and Geminis preferred."[4] Astrological predictions, in one form or another, seem to be a major influence on many people in many places.

Why is this so? One clue might be provided by a recent study that suggests that those who believe in astrology also tend to believe more in the importance of "obedience."[5] This illustrates that astrology, like so many other superstitions and belief systems, is a way of trying to tame and control the unpredictable and non-obeying randomness in all of our lives.

Of course, with astrology as with so much else, the big question is, is it true?

Astrological Predictions?

For many people, horoscopes are mere harmless fun, read for amusement only. Others might be inspired by them to think about important issues in their lives. Still others might take comfort from their predictions, or use them to help make decisions. (This is sort of like that old trick: if you're torn between two choices, flip a coin to decide—and then if you're disappointed with the result, you know you should really do the opposite.) These uses are all reasonable, and I have no objection to any of them.

On the other hand, many aspects of astrology imply that astrologers are making actual predictions, or have some special knowledge of our daily lives. They classify everyone into one of 12 different zodiac signs, based on their birthday, and provide specific advice and forecasts for each sign. If valid, this is a very effective way to tame luck—read your horoscope and decide what the day has in store for you. But is it really so?

I certainly wish it were. Who wouldn't want to be able to figure out their future—and know what to embrace, what to avoid, what to decide—all by considering birthdays and following the movements of planets? The appeal is clear, and the temptations strong. But what of the science?

On the one hand, astrology has its roots in genuine astronomical science, and does (in some cases) involve careful and sophisticated tracking of the motions of the planets and so on. But on the other hand, how could these planetary motions possibly influence such things as personalities, daily events, and luck?

Some have suggested that gravity could play a role in this influence. However, the gravitational force from planets is about as insignificant as the gravitational force exerted on a mother by a doctor standing beside her while she gives birth.[6] No one thinks that a baby's personality is affected by the tiny gravitational attraction towards whoever happens to be standing in the room when the baby emerges, so how could the gravity from other *planets* play a role? And magnetic influences from the planets are similarly minuscule. So, what planetary force could possibly cause the effects that astrologers claim?

Of course, it is conceivable that there could be some other force at work, besides gravity, that science has not yet discovered, but which astrologers have tapped into. Indeed, astrologers are quick to point out that science does sometimes recognize a scientific phenomenon before understanding its cause.[7] For example, we have already seen how Ignaz Semmelweis demonstrated the usefulness of handwashing in preventing the spread of disease well before Louis Pasteur established the existence of germs. Some scientists were indeed skeptical of Semmelweis's claims, but experiments replicated later proved that he was correct. So we should not dismiss astrology simply because its mechanism is not understood.

Furthermore, it has been established that *some* aspects of birth timing do indeed play a role in future development. For example, studies indicate that the month in which babies are born can have an impact on what diseases they will contract.[8] This result is surprising, but it makes a certain sense. Indeed, the babies studied were from the northern hemisphere (New York), so the month they were born probably affected how old they were when they faced their first winter, how often they were taken for outdoor walks as an infant, how brightly the sun shone when they first learned to see, and so on.

Similarly, it is possible that personalities are affected by birth order—whether an individual is the oldest, or middle, or youngest child in their family. Some early studies about this issue were rightly criticized for not taking into account family size: smaller families have fewer children but still just one oldest child, and therefore a greater fraction of oldest children, and so on, which skews the results. However, more recent studies do seem to suggest that birth order has at least some small effect on personality; for example, individuals are slightly more likely to marry people at the same birth rank as themselves.[9] Again, this is interesting but not shocking, since birth order tends to have a significant effect on an individual's experiences while growing up.

But a mysterious unknown force from other planets? Sounds farfetched. Nevertheless, to be scientific, further investigation seems called for. How should I proceed?

Evaluating My Horoscope

For starters, while writing this, I looked up my horoscope for today in my local newspaper.[10] Under my star sign (Libra), it declared, "You might feel as if you are on top of a project, at least until a close associate tries to challenge the very basis of your thinking. Discuss this person's perspective, and listen to his or her suggestions. Avoid getting defensive. Think positively." So, what am I to make of this?

Well, some of this horoscope consists simply of advice, and not actual predictions. Now, I do agree that discussing others' perspectives, listening well, not getting defensive, and thinking positively are all very wise bits of advice that should be followed by nearly everyone (not just Libras) nearly all of the time. So I support and appreciate the horoscope's wise counsel. However, it does not provide evidence of any special insight or knowledge about my own personal circumstances.

The horoscope does also make one actual prediction—namely that an associate will challenge my thinking. Well, first of all, this prediction is quite general. Most days for most people include *some* sort of associate challenging us in *some* way. So this is the sort of description that will apply, in some sense, to most people on most days. In terms of our luck traps, it is an *extra-large target*. It follows that many people might declare this prediction to have come true for them. But that proves nothing, since it would also tend to come true on most other days, and for most other people (of any star sign) as well.

Yet, as it happens, this prediction did not apply to my day at all! I spent my entire day today working on writing this book, at home. I did not discuss my writing with anyone, so there was no "challenge" from any "associate"—not today, anyway. I can therefore declare with confidence that, in my case at least, this horoscope's prediction was completely false.

Related to this, consider an amusing story about horoscope writing by the magician and skeptic James Randi.[11] At age 17, he agreed to write an astrology column for his friends' newspaper. He created his horoscopes by rearranging pages from old astrology magazines in a completely random way. Nevertheless he witnessed readers who "squealed with delight" at the accuracy of his forecasts and pronounced them as being "right smack on." This further illustrates that many astrological predictions are so general—such *extra-large targets*—that almost anyone of any star sign could find some truth in them.

Astrological Evidence

A few examples and anecdotes don't prove very much. What is needed to draw conclusions are large-scale controlled studies. Are there any such studies? Yes, there are.

One important experiment along these lines was published by Shawn Carlson in 1985 in the prestigious journal *Nature*.[12] He recruited 28 respected astrologers (nominated by the National

Council for Geocosmic Research, a leading astrology organization), to study 128 subjects. Each astrologer was shown a randomly chosen subject's natal chart (a horoscope based on the place and time of the individual's birth), together with three detailed personality profiles—one describing the subject, plus two describing other "control" individuals. The question was, what fraction of the time could the astrologer correctly determine which profile corresponded to the subject under investigation? Random guessing would lead to about one-third (33.3 percent) correct, but the astrologers expressed confidence that they would guess correctly over half the time. What happened? The astrologers' prediction accuracy averaged 34 percent (with a standard deviation of 4.4 percent),[13] entirely consistent with the one-third random-guessing hypothesis. In short, this experiment suggested that respected astrologers could not identify which horoscope corresponded to which person any better than they could by guessing randomly. This is quite a serious indictment of astrology.

Another study looked at the distribution of zodiac signs among people in the science profession.[14] The birthdays of 14,662 scientists were obtained and classified according to their astrological sign. (Birthdays at the transition from one sign to the next were excluded.) The counts by star sign were:

Aquarius	1,217
Pisces	1,193
Aries	1,158
Taurus	1,185
Gemini	1,153
Cancer	1,245
Leo	1,263
Virgo	1,292
Libra	1,267

Scorpio	1,246
Sagittarius	1,202
Capricorn	1,241

These counts are all quite close to the average (1,222),[15] suggesting that astrological signs have no effect on whether or not someone becomes a scientist. Some astrology websites claim that Capricorn, Aquarius, and Scorpio are best suited to science.[16] However, a look at these numbers shows that the count for Aquarius is below the average value of 1,222, while those for Capricorn and Scorpio are marginally above but well within the expected range. So this experiment suggests that star signs do not help to predict scientific ability at all.

In a similar spirit, another study looked at the effect of astrological sign on suicides,[17] which surely have a link to personality type. They examined the records of all 502 suicides in one English county in a 12-year period and determined the star sign of each victim. The totals were:

Aquarius	44
Pisces	36
Aries	39
Taurus	53
Gemini	48
Cancer	36
Leo	48
Virgo	51
Libra	38
Scorpio	38
Sagittarius	30
Capricorn	41

These counts show some variation, but well within what would be expected by random chance alone.[18] To be fair, after much searching, the study did find some anomalous results, such as more suicides by hanging among Virgos than others. But as we have seen, with so many possibilities to study (many different suicide methods, 12 different possible signs, etc.), it is not unexpected that some result stands out by chance alone due to the *shotgun effect*. Overall, this study seems to further refute the idea that astrological sign affects personality.

Several other studies are also available. One, published in 1974 in the *Journal of Psychology*, studied 130 students, looking at their self-ratings (and their friends' ratings of them) on various personality characteristics (aggressive, ambitious, creative, extroverted, and intuitive).[19] It compared these subjective ratings to those that would be predicted by combining their sun and moon and rising zodiac signs. The authors found no significant correlation, and hence no evidence to support astrology's predictions.

Another study, in 2008 in the *Journal of General Psychology*, generated two different personality profiles for each of 52 college students.[20] One of the profiles was based on astrological predictions from their true birth date and astrological information, using a computer program endorsed by astrologers for this purpose. The other was a randomly generated bogus decoy. The students were asked to choose which of the two profiles accurately described them. If they guessed randomly, they would be correct only half the time. So how did they do? The study found that only 24 of the 52 students chose the profile based on their real birth date. This is just 46.2 percent—slightly less than the half correct that would be expected by luck alone.

A 1980 study considered the relationship between birth date and college major for 10,313 university graduates.[21] It divided the graduates into 13 general areas of study (humanities, music, commerce, science, etc.). Within each area, separately for male and female students, they counted the number of graduates born under each of the 12 star signs. In 25 of the 26 cases, they found no significant relationship. The one exception was female medical students, where they found a slightly higher-than-expected number of Aries (apparently consistent with astrological predictions) and Capricorns (apparently the opposite of astrological predictions).[22] These findings seem quite consistent with random chance. Indeed, the slight Aries agreement is surely a *shotgun effect*, because of all the possible combinations of different study area and gender and star sign that were considered.

Another study, in 1996, looked at the relationship between personality and sun and moon and ascendant zodiac signs among 190 university students.[23] It performed numerous tests, involving different positions of each of the signs and multiple different personality scores. It found no significant correlations in any of these tests, except for one. Namely, subjects with both sun and moon "positive" (or "masculine") signs were more extroverted than those with both signs "negative" (or "feminine"). Again, this one significant correlation was presumably due to luck alone, another *shotgun effect* from considering so many different tests arising from so many possible choices.

Finally, a large-scale study in 2006 examined 4,462 veterans of the Vietnam War, together with 11,448 young adults from a longitudinal study.[24] With this large database, they considered numerous potential correlations regarding four personality traits (psychoticism, extraversion, neuroticism, and social desirability), together with sun signs, either individually or grouped by "element" (fire, water, air, earth) or "gender" (masculine/feminine). Out of all these tests, how many statistically significant relationships did they find? None at all. We don't need a *shotgun effect* to explain *that* result!

The bottom line is that a number of people have conducted studies in search of evidence that astrological predictions really do come true more often than they would just by luck alone. And in each case they found no clear evidence at all to support astrology.

Astrological Response

How have astrologers responded to such scientific studies, which apparently show no causal effect from astrology? Well, sometimes they have argued that astrology should not be subject to scientific investigation. For example, one astrologer wrote that "the practice of astrology by most astrologers is better defined as an art or a craft than as a science and it would be wrong for these type of astrologers to claim to be scientists. As such it would also be equally wrong for a scientist who has not studied astrology to consider him or herself qualified to judge such practices since they are outside the realm of science." He decried the "quantitative tests when the data requires qualitative analysis that would be better addressed by those who understand astrology."[25]

I am not completely unsympathetic to this viewpoint. I'm sure that many detailed astrological "readings" involve much discussion and back-and-forth that is tailored to the responses of the client, and they may indeed end up helping the client as a form of therapy. And I have no problem with that at all. But then the question becomes: Is this therapy actually aided by the readings of the positions of the planets and so on?

Testing this properly wouldn't be easy. Therapy and life assistance are so subtle and so individual that evaluating them precisely is a challenge. In addition, astrological readings could well have a *placebo effect*, where clients take therapeutic advice more seriously simply because they *believe* it comes from a higher power rather than from a lowly therapist, even if the astrological readings themselves are actually meaningless. A proper test would have to involve astrologers conducting complete and detailed sessions with clients, sometimes based on their true birth information, but other times based (unknown to them) on falsehoods (such as the wrong birthday). Then, the helpfulness level of the sessions would have to be measured somehow. At that point, if the sessions using true birth information were consistently and significantly more helpful than those using false information, it would provide evidence in favour of astrology. By contrast, if there were no difference between them, this would provide evidence against the validity of astrological predictions.

I don't know if any controlled scientific tests of this type have been conducted. But as we have seen, tests of simpler astrological predictions involving sun signs have failed to find any clear effect. In summary, there is no clear evidence that the planets actually influence people's behaviour or destiny at all. The reality is that so many individuals accept the validity of astrological therapy despite the absence of any convincing evidence.

At other times, astrologers question the motivation and the methodology of these scientific studies and argue that other "studies" of their own are more reliable and actually demonstrate astrology's true power. Some of these objections are passionately argued.[26] In some cases they may even cast doubts on *some* aspects of the original studies—not enough to invalidate the studies, in my opinion, but enough to cause sufficient confusion to allow astrology defenders to remain committed to their practice.

Related to this, for 19 years the James Randi Educational Foundation offered a $1 million prize (since terminated) to anyone who could produce evidence of paranormal abilities (including astrological predictions) under controlled conditions.[27] Many people tried to claim this prize,[28] but they all failed under proper examination, and no one ever received a single penny.[29]

In response, one astrologer complained that the test was unfair: Randi was requiring a preliminary test with a p-value of less than one chance in 100 (1 percent), and then a careful replication with a p-value of one chance in 100,000 (0.001 percent). Since 100 times 100,000 equals ten million, they complained that "a winner is still required to outperform odds of up to one in ten million" and huffed, "You are ten times more likely to be struck by lightning in any one year than beating the Randi odds!"[30] What they were forgetting, of course, is that the p-value represents the probability of obtaining a significant result just by pure, dumb luck, when there is no true effect at all. So Randi's requirements concern the chance of having to pay a million dollars due solely to a *lucky shot*. If an effect were actually real, then with enough testing it should be able to overcome any standard of proof and reach any required p-value level—yes, even one chance in ten million. And no such success has ever been recorded.

However, with so many people still believing in astrology, it seems that these sorts of statistical analyses and scientific tests and published studies will not convince them. So, I wondered, what more could I do to investigate this issue further? To try to cut through the confusion, I decided to do some tests of my own.

Astrological Professions?

For starters, many astrologers claim that your astrological sign can affect which careers you are best suited for. But is it true? And, how could I test this myself?

To keep it simple, I decided to stick to the astrological (sun) signs used in the published horoscopes that we are all familiar with. To test their influence, I would need to find lists of birthdays of people in different professions. From a quick web search, I determined that politicians' birthdays are often available online, so I decided to investigate them. I quickly found a 2014 article about astrology that listed the birthdays and star signs of all members of the US Congress.[31] I also searched for astrology websites and found just four that addressed this question, each of which claimed that people born under Aries are best suited to being a politician (though one of them mentioned three other signs too).[32]

Now it was time for a test! I was actually nervous as I had my computer count up the star signs for me. If there was an excess of Aries in the list, how would I react? I would first have to do a statistical test to see if the excess was statistically significant or was just luck. If the latter, I would have to explain that in a sufficiently convincing way to satisfy any doubters. If it *were* significant, then that alone wouldn't convince me that astrology was real, but I would have to admit that my first test showed some support for astrology, and I would then have to figure out how to investigate the question in more depth.

With some trepidation, I looked at the results. And what did I find? It turned out that there were actually slightly *fewer* Aries than would be expected by chance. Specifically, a total of 531 members of the House of Representatives and US Senate were listed. If each of the 12 star signs were equally likely to be represented, then we would expect about 531/12, or 44.25 of them, to be Aries. In fact, according to the list, there were 41 Aries in Congress. This is close to 44.25, and well within the margin of random error. But it is a bit less than expected, and certainly not significantly more, as the astrologers claimed. A clear, unambiguous defeat for the astrological predictions.

Incidentally, the article that listed the politicians' star signs also quoted an astrologer as saying "we can expect big things from Capricorn Sens. Rand Paul and Ted Cruz . . . [The astrologer] likes the prospects of Minnesota Democrat Amy Klobuchar and Republicans Marco Rubio and Mike Lee, all Geminis. Rep. Paul Ryan and Sen. Tammy Baldwin are also primed to have lasting impact." Since then, Klobuchar and Lee and Ryan and Baldwin have all remained in their same positions, without change, so that part is a tie. But Paul, Cruz, and Rubio all ran for the Republican presidential nomination the following year, and all lost. Astrology fails again.

Astrological Nurses?

To do a larger-scale test, I considered the field of nursing. Of the seven astrology websites that I found that commented on the best signs for nurses, six asserted that Pisces are best suited to being nurses,[33] while three each mentioned Taurus and Cancer. To test the claim, I obtained, through a contact,[34] an anonymized list of the birthdays of all of the 158,077 nurses currently practising in Ontario. The question was: Are these nurses significantly more likely to be Pisces than could be explained by random chance?

I wrote a computer program to count the number of nurses of each sign, based on the data I had received. The results were:

Aquarius	12,660
Pisces	13,186
Aries	13,070
Taurus	14,128
Gemini	13,672
Cancer	13,892
Leo	13,183
Virgo	13,690
Libra	13,308
Scorpio	12,801
Sagittarius	12,473
Capricorn	12,014

What did this prove? Well, the average here is about 13,173 per sign. And the Pisces count was 13,186—almost exactly the average, and well within the usual random variation. So this experiment indicates that, contrary to what many astrologers claimed, Pisces are not any more likely than others to become nurses. (On the positive side, I suppose this shows that at least people aren't pursuing nursing in opposition to their inclinations, simply because some horoscope told them to.)

The numbers of nurses under the other signs are all fairly close to the average, but not dead on. In particular, the figures for Taurus and Cancer are both a little bit higher than would be expected by chance alone. Now, Taurus and Cancer were both mentioned as recommended signs for nurses on *some* of the websites I checked (though much less often than Pisces was). So what was I to make of this surprising result?

I decided to check for an *alternate cause*. Specifically, births are slightly more common at some times of the year than others, so perhaps the entire population—not just nurses—also has a somewhat uneven distribution of signs. To understand this situation better, I obtained counts of the total numbers of all births in Ontario on each date for one complete year.[35] I then wrote a computer program to match the birth dates to their corresponding astrological sun signs. The counts were as follows:

Aquarius	11,215
Pisces	11,515
Aries	11,590
Taurus	11,770
Gemini	12,165
Cancer	12,865
Leo	12,480
Virgo	12,810
Libra	12,010
Scorpio	11,500
Sagittarius	11,300
Capricorn	10,380

The average of these counts is 11,800. For the general population, then, the number of Pisces is a bit below average, the number of Taurus is nearly average, and the number of babies born under Cancer is significantly above average (presumably because summer births are more common).

This can all be seen more clearly in the following graph. Here, the solid curve with circles is the percentage among the general population, the dashed curve with squares is the percentage among nurses,

and the straight line is what we would expect if all signs were exactly
equally likely:

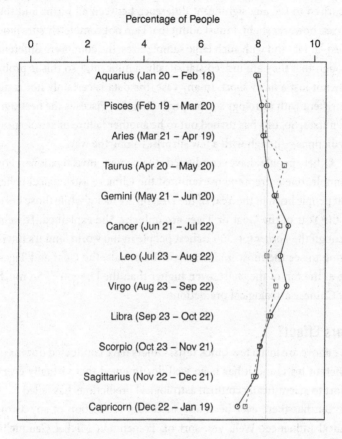

Percentage of People

This graph tells us that the percentages of Pisces among nurses,
and among all births, are both very close to (and just slightly less than)
the expected value of one person in 12 (about 8.5 percent). For Taurus,
the percentage among all births is right on the average, but the percentage of nurses is somewhat higher. For Cancer, the percentage of
nurses is a bit higher than average, but the percentage of overall births
is even higher than that. So, the data suggests that, in Ontario at least,

people born under the sign of Taurus are indeed a bit more likely to become nurses, but Pisces and Cancer show no such tendency. I was surprised to see any significant difference between all births and the nurses, however slight. I must admit that I am not completely sure how to explain it (and with such large sample sizes, the counts are different enough that they are indeed statistically significant,[36] so this is probably not just a *lucky shot*). In any case, the data is certainly not at all consistent with astrology's main prediction that Pisces is the best sign for nurses. So, this has turned out to be another failure of astrological predictions—though with a few surprises along the way.

Other people have conducted their own investigations. For example, one entrepreneur examined the Chinese astrological belief that people born in the Year of the Dragon are lucky, while those born in the Year of the Goat or Tiger are unlucky. She explained, "I went through the *Forbes* top 300 richest people in the world, and it's interesting to see the most undesirable two animals, the Goat and Tiger, are at the top of the chart, even higher than the Dragon."[37] So much for Chinese astrological predictions.

Mars Effect?

The above are just a few quick tests. Others have conducted dozens of different tests, and it has been well documented that virtually every effort to scientifically confirm astrological predictions has failed.[38]

But has there ever been any successful replication of any astrological influence? Well, yes, sort of. Frenchman Michel Gauquelin and his wife, Françoise, spent many years testing all sorts of astrological claims about relationships between human characteristics and achievements, and the positions of the planets and stars when people were born.[39] Their results were consistently negative, disproving claim after claim. With one exception.

Namely, the Gauquelins found a "Mars effect," in which certain people of exceptional achievement in sports and the military were slightly more likely to be born at a time soon after the planet Mars is seen to either "rise" (from the perspective of their birthplace) or be at the midpoint between rise and fall (its "midheaven"). Gauquelin claimed that this relationship was statistically significant, and that it not only held in France (where he originally studied it), but was successfully replicated in other countries too. And unlike many astrologers, he provided detailed data about his claim, which he allowed others to verify.

So, what are we to make of this finding? Were there any luck traps at play here?

There certainly were. Most obviously, there is a huge *shotgun effect*. There are many other planets that could have been considered instead of Mars. And there are many other aspects of the planets that could have been considered besides rising and midheaven, including the more usual astrological fact of their position relative to the star constellations (which corresponds to day of the year rather than time of day). Plus, there are many different types of careers and many ways of being "exceptional" in them. Gauquelin had tried out many combinations before settling on those few that produced positive results.

As one concrete example, Gauquelin divided famous athletes into two groups: those he decided were "iron-willed," and those who were "weak-willed" and succeeded with minimal effort only because of built-in physical attributes. He found that the Mars effect applied only to the first group. Obviously, this raises questions of how he determined which performers belonged in each group, while providing another *shotgun effect* in terms of which group of famous athletes we should or should not consider.

In another twist, when it came to identifying exceptional *scientists*, then suddenly it was not Mars that predicted them, but the position of *Saturn*. This was clearly another *shotgun effect* at work: with so many combinations available, some correlation was likely to turn up by luck alone.

Furthermore, the claimed Mars effects were quite small. For example, in his study of 3,438 French military leaders, Gauquelin found that 680 of them (19.8 percent) were born with Mars in an appropriate position, compared to 16.7 percent of the population overall. In other words, this claimed astrological effect, if in fact it does exist, only influences about 3 percent of these "exceptional" people—quite a small influence indeed.

In addition to all of this, many scientists criticized Gauquelin's methodology,[40] largely on the grounds of bias in precisely which individuals did or did not get counted as "exceptional," thus leading to a form of *many tries* luck trap. Furthermore, a group of skeptics published an article claiming the Mars effect did not hold,[41] but one of their members later accused them of covering up some evidence,[42] which certainly did not help to resolve the controversy.

We should always be hesitant to dismiss evidence that has apparently been replicated and shown to be statistically significant. However, this Mars effect appears to be a very small influence at best, and one whose verification is subject to numerous luck traps. It seems woefully insufficient—especially in the face of so many other failed astrology experiments—to provide any clear evidence of astrology's predictive powers.

Numbers and Numerology

What about the related field of numerology? One version of numerology involves trying to predict people's personalities and careers based on their specific date of birth. Now, on the one hand, I am always in favour of using numbers and numerical reasoning to better understand the world. (I couldn't help but feel a burst of pride when the crime-fighting mathematician Charlie Epps, on the television show *Numb3rs*, boldly declared that "everything is numbers.")[43] On the other hand, I am inclined to be skeptical that any clear link exists between a precise birth date and later personality traits.

How could I test this? Well, one magazine generously provides a free online numerology service.[44] So I eagerly typed my birth date into their "Numerology: What's your career path?" tool and immediately received an automated reply that told me such gems as: I have determination and ambition, I am active and creative, I shouldn't be afraid to demonstrate what I'm capable of, I should be sure to aim high, and so on. What can we make of this? Well, first of all, these claims are all very general and would apply to lots of people in lots of situations, thus giving the numerologists *many tries* to succeed. Indeed, who among us doesn't have (or think we have) at least some determination and ambition? Who isn't (at least sometimes) active or creative? Who shouldn't try to "demonstrate what you're capable of" or aim high? It's hard to imagine more of an *extra-large target* than this.

On top of all that, my numerological forecast still wasn't very accurate! In addition to the above generalities, it also declared that I "could lead an expanding company" and suggested such careers as "business lawyer," "editor-in-chief," "site manager," "marketing director," and "designer," none of which applies to me at all. If their long list of proposed professions was a form of *shotgun effect*, then most of the lead completely missed the target.

Well, there is one possible exception. They did say that I would also excel at being a "writer." Gosh. I suppose that, now that you have read most of this book, it's up to you to evaluate *that* particular claim for yourself!

Astrological Reflections

I wish I could test the Mars effect myself, on some fresh evidence—say, on exceptional achievers in Canada. However, in addition to complicated calculations about the position of Mars, it requires knowing the exact time and place of birth, data the Gauquelins obtained in some countries, but which is not easily available in others. So, a test would be challenging to carry out.

If I were able to do such studies, I am reasonably confident that the Mars effect would not be replicated—in other words, exceptional Canadian achievers would show approximately the same distribution, in terms of the position of Mars, as would the general population. But I am even more confident that even if I did so, and even if my results showed no Mars effect at all, proponents of astrology would continue to believe that planets influence our personalities—with or without sufficient evidence.

Indeed, people who wish to believe in astrology's influence will find justifications any way they can. For example, one astrologer found that of the first 43 US presidents, the most common astrological sign was Cancer. (No, not Aries, which those astrology websites had claimed was the sign best suited to politicians.) They then "explained" their finding as follows: "It turns out that the United States of America has a Cancer sun sign, with a small stellium in Cancer. (U.S.A. was officially born on July 4, 1776, Philadelphia, PA.) Given this information, it's not surprising that America loves presidents with a strong Cancer influence."[45] I should not have to tell you by now that such incredibly flexible interpretations give them *many tries* indeed to justify their findings.

Such weak arguments, with so many potential luck traps, certainly do not qualify as the extraordinary evidence that would be required to truly demonstrate an unexplainable influence of birth dates or planetary positions on human achievement.

CHAPTER 21

Mind over Matter?

The human brain is an incredible organism. How could a few pounds of squishy goo be capable of solving puzzles, forming friendships, writing novels, making decisions, falling in love, and so many other incredible cognitive functions? It truly boggles the mind.

Brains have other surprising features. They can get fooled by visual tricks.[1] They can forget things. They can be sure something happened when it never did. And they can compose intricate dreams that feel completely real.

Perhaps because of the brain's incredible internal functions, many people believe—or wish to believe—that the brain also has external powers. They think the brain can telepathically move matter, or can learn about the outside world not through the usual senses—sight, hearing, touch, smell, and taste—but through some sort of direct brain connection, or "extrasensory perception."

It certainly happens in fiction. The *Star Wars* movie franchise may have its spaceships and laser blasters and explosions, but it is best known for "the Force," a mysterious power through which well-trained disciples can somehow detect the presence of other people, confuse weak minds, deflect weapon fire, and choke enemies. The kindest words to speak in the *Star Wars* universe are not "take care," or "best wishes," or even "good luck," but rather "may the Force be with you."

Can such special brain powers arise in real life? Many people think so.

Psyching the Psychics

Peter Popoff called himself a "faith healer." He led large meetings in which he would approach audience members and claim that he could, through the power of God, heal their various afflictions (cancer, needing a walker, etc.) and make them healthy again. What made him seem convincing was that when he first approached an audience member, he announced their name and affliction, as well as personal details like their home address and relatives' names. How could he know all this information, if not through divine powers?

Not surprisingly, the skeptic James Randi was skeptical. Looking carefully at the videos of Popoff's performances, he noticed a small earpiece in Popoff's ear and immediately became suspicious. Did Popoff require a hearing aid? Or was he being given information through decidedly non-divine methods?

Randi got a technician to design a radio scanner that was able to pick up radio signals on their way to Popoff's ear. It quickly became clear that Popoff's wife, Elizabeth, was speaking to him through a wireless radio transmitter during his performances. By reading the "prayer cards" that audience members had been asked to fill in before the show, Elizabeth was able to give Popoff precise information about each audience member's name, affliction, address, and more. So Popoff was obtaining all of his information not from God or other supernatural means, but from his wife through his earpiece. This was not a misunderstanding, or a biased experiment, or a luck trap. It was outright fraud.

Randi cleverly created and broadcast a video of Popoff's performance, together with the intercepted radio signals, thus demonstrating the truth of the matter to everyone.[2] Popoff was disgraced, and he declared bankruptcy the following year. Justice was done, albeit after Popoff had already collected millions of dollars from the people he had duped.

Incredibly, that is not the end of his story. A few years later, Popoff made a comeback, convincing a new generation of his divine healing powers. Once again, just like in fiction, many people wanted so badly to believe in special supernatural powers that they were willing to overlook clear evidence of fraud.

Randi exposed self-proclaimed "psychics" in other ways too. For example, his "Project Alpha" involved recruiting and training two young magicians to *fake* psychic powers.[3] They were so convincing that they were able to fool serious scientists into thinking they had true psychic powers, even after conducting detailed "controlled" scientific experiments for more than 160 hours over four years. They used simple tricks like switching labels on objects after they had been measured, to make it look as though they had altered their size and physical characteristics. Those scientists even presented videotape of the experimental sessions at a convention of the Parapsychological Association as proof of the validity of psychic powers—only to learn later that their "psychic subjects" were nothing more than magicians performing tricks. In other words, the apparently psychic phenomena in their videotapes had an *alternate cause*, and there was nothing supernatural about them at all.

Cold Readings

Fraud does not have to be as blatant as Popoff's to be effective. One of the oldest tricks for appearing "psychic" is to make assertions that are so general that they would apply to most people. Like with astrological predictions, the key is to create an *extra-large target* that is much easier to hit. Classic "stock spiel" phrases include "You get along well without effort"; "You are neither overly conventional nor overly individualistic"; "You are adaptable to social situations"; "Your interests are wide"; "You have found it unwise to be too frank in revealing yourself to others"; "You prefer a certain amount of change and variety"; and "You have a tendency to be critical of yourself."[4] They sound insightful, but the reality is that most people would agree with most of them. A psychic reading that contains such phrases will generally be judged to indicate special knowledge of the subject's unique personality—even though it actually indicates no such thing.

A classic experiment by psychologist Bertram Forer back in 1949 nicely illustrated this phenomenon. Each of 39 students in a class was asked to fill in a questionnaire about their hobbies, reading materials, ambitions, etc. A week later, in response, each student was handed a typed "personality description" and asked to rate, on a scale from zero to 5, the extent to which it revealed basic characteristics of their personality. Of the 39 students, 16 gave a rating of 5 out of 5; 18 gave a rating of 4; 4 gave a rating of 3; and the final student gave a rating of 2. The mean rating was thus 4.26 out of 5, or 85 percent—quite high indeed.[5]

Did this demonstrate that a trained psychologist was able to use questionnaire answers to derive important personality insights? Hardly. Each of the 39 students had received an *identical* personality description, consisting of assorted stock phrases similar to those mentioned above (his were taken largely from an astrology book), including "Disciplined and self-controlled outside, you tend to be worrisome and insecure inside"; "Some of your aspirations tend to be pretty unrealistic"; "Security is one of your major goals in life," and so on. Once again, sufficiently general statements are made to seem insightful—even if, in reality, the same assessment was given to everyone—thus creating an *extra-large target* and virtually guaranteeing success. This tendency is so strong that it is sometimes called the Forer effect.

Similar comments apply to fortune cookies, whose predictions (when they make any at all) are often so general that any sufficiently convinced person can find some truth in them. This is nicely summed up by one satirical image that shows, emerging from a broken fortune cookie, the bold prediction, "You will continue to interpret vague statements as uniquely meaningful."[6] Which just about says it all.

Seeing the Future?

Most scientists, including most psychologists, do not believe there is any evidence at all for any kind of extrasensory perception. However, there are exceptions. In particular, a Cornell University psychologist named Daryl Bem caused quite a stir when he published a paper claiming that he had experimental evidence for "precognition," the ability of subjects to sense events before they happen.[7]

For example, Bem showed subjects two different curtain images on a computer screen. The subjects were asked to click on whichever curtain they thought had a picture "behind" it, at which point the curtain would open and show them if their guess was correct. What did he find? For "non-erotic" pictures, the subjects guessed correctly 49.8 percent of the time—just below the expected 50 percent, and well within the usual margin of random error. For "erotic" pictures, he initially found that women got more than 50 percent correct but men did not, so he switched to "stronger and more explicit images from Internet sites for the men." (These various adjustments already seem to run the risk of introducing a *many tries* luck trap.)

After much experimenting, Bem eventually found that his subjects selected the correct curtain 53.1 percent of the time.[8] This is only slightly above the 50 percent that one would expect by chance alone, but he claimed it was still statistically significant. Furthermore, he claimed he had carefully avoided such luck traps as *biased observation* and *false reporting*. Thus, he asserted, the (small) effect that remained must be the result of genuine precognition—the ability of subjects to sense, occasionally, events that had yet to happen. He also included some other experiments, including two about the related question of whether subjects can better recall words from a list if they review and practise the words afterwards, and again claimed statistically significant results.

Bem's experiments were sufficient to convince the journal editors to publish his article. They admitted in an accompanying editorial that Bem's "reported findings conflict with our own beliefs" and were "extremely puzzling" and "strange," but that it was their "hope and expectation that the current paper will stimulate further discussion, attempts at replication, and critical further thoughts."[9] And so Bem's article was published, and widely reported in the media, to great surprise and much skepticism.

Quantum Magic?

Most scientists have a very solid reason for not believing in precognition: it seems to be scientifically impossible. According to classical science, governed by Newton's laws of motion,[10] the universe is completely deterministic.[11] Objects and forces and actions at one instant directly cause outcomes at the next instant, and on and on, without any uncertainty or deviation. In short, the future is caused by the past. This means there is just no way for a future event to cause something else to happen now—it just doesn't work that way. So the future location of an image cannot affect the location a subject guesses right now. And studying words in the future cannot help a subject to recall the words right now. The sorts of results that Bem claims are impossible, plain and simple.

I mostly subscribe to that view, and it is one of the reasons I am very skeptical of Bem's claims. However, the situation isn't quite as clear-cut as that, because of two post-Newtonian scientific developments. The first is Einstein's general theory of relativity, which might just possibly make it possible to travel backwards in time, either via space-time wormholes or by achieving faster-than-light travel.[12] If so, then conceivably this could allow someone to figure out at some future time where an image is, and then go back in time to choose the correct location now.

Might *that* provide us with an explanation for Bem's results? Well, I certainly love time travel in fiction, from H.G. Wells's classic novel *The Time Machine* to all those *Star Trek* stories, right up to the Audrey Niffenegger book (and later film) *The Time Traveler's Wife* and beyond. As an explanation for precognitive behaviour in psychology experiments, however, it sounds extremely far-fetched. But I can't quite call it impossible.

The second development is the theory of *quantum mechanics*, which says that at the extremely small sizes of atoms and molecules, classical science breaks down. Locations of particles are no longer absolute; rather, they are random with various probabilities.[13] Energy isn't always conserved; rather, it can change slightly over brief time periods, which then allows for unusual effects like "quantum tunnelling."[14] Particles can become "entangled" so that they influence each other even when they are far apart.[15] And, most important for us, quantum mechanics is no longer deterministic—those random probabilities mean that the future is no longer completely determined by the past. In particular, this just might lead to the possibility of a concept called "retrocausality," where effects can occur before their causes, meaning the future can affect the present after all.[16] Furthermore, some scientists (though probably not too many) now believe that quantum mechanics may be the key to understanding the mystery of human consciousness.[17] If so, then maybe—just maybe—quantum effects could perhaps allow future events later on (like a later photo, or later word practice) to influence the human brain's knowledge and actions right now (like choosing the correct image, or remembering a word). This seems extremely far-fetched, and I doubt that such an effect is possible. But can we really know for sure?

In Douglas Adams's classic book series *The Hitchhiker's Guide to the Galaxy*, years of study convince the editors of the guide to upgrade the entry about the planet Earth from "harmless" to "mostly harmless." In a similar spirit, I would say that the theories about relativity and quantum mechanics compel me to upgrade my initial assessment of Bem's results from "impossible" to "almost impossible"—I don't buy them for a second, but perhaps they shouldn't be completely dismissed out of hand. Instead, they should be studied further. But how?

Seeing the Replication?

Had Bem discovered evidence of a special precognitive ability? His experiments seemed to indicate a small effect. But perhaps this was just luck, or (even worse) some overeager lab assistant somewhere doctoring a few of the results. The only way to be sure was to—that's right—try to replicate his experiments. And indeed, other psychologists soon did. In many cases, their replications did not succeed.

For example, a Swedish mental health researcher attempted to replicate Bem's experiment about the picture selection in three ways, and found success rates of 50.0 percent, 47.2 percent, and 50.8 percent, none of which was significantly above the 50 percent expected by chance alone.[18] A team of researchers led by a Carnegie Mellon University business professor tried to replicate Bem's other two experiments, about subjects better recalling words from a list if they review and practise the words afterwards.[19] Those researchers found that in four of their seven experiments, participants actually recalled more control words than practice words—the opposite of Bem's claims. So, no confirmation of Bem's results there.

A third group, led by a University of Edinburgh psychology professor, also tried and failed to replicate Bem's findings about improving memory through post-test practice.[20] Interestingly, the journal that had published Bem's findings, the *Journal of Personality and Social Psychology*, apparently refused to publish this replication attempt, saying that the journal's policy is to avoid publishing mere attempts to replicate something already discovered.[21] This is quite problematic since, as we have seen, replication is the best (indeed, often the only) way to determine whether a result is accurate or is just luck. As a result, many psychologists have been calling for more emphasis and value to be placed on the publication of such replications.[22]

Of course, there are many reasons why a psychological experiment might fail to be replicated. But in this case, it seems that the most likely explanation is that there was no true effect in the first place. To be fair, in a follow-up paper, Bem and others conducted a "meta-analysis" of various precognition experiments and asserted that, taken together, they still provided evidence of an effect.[23] But the methodology for this meta-analysis has also been criticized,[24] so at the very least, uncertainty remains.

I would love to conduct my own experiment about this, by setting up a web page allowing people everywhere to guess picture locations to their heart's content. Indeed, it would be quite easy to program such a site. But there are challenges with this plan, due to Bem's requirement that only sufficiently "erotic" pictures can be used. Indeed, the main picture source that he recommends (the International Affective Picture System, or IAPS[25]) explicitly prohibits placing its images "on the internet in any form,"[26] which seems to rule out my plan—to say nothing of Bem's later requirement about "stronger and more explicit images." Of course, my experiment could instead use images that I find myself on the internet (if I could bear to look for them)—but then Bem might object that my images lacked whatever crucial property is required to generate precognition. So, what to do?

Nevertheless, I'm reasonably confident that if I ever *did* conduct proper experiments, then just like those other attempts mentioned above, my experiments would *not* replicate Bem's findings and would thus provide further evidence against the existence of precognition. Perhaps one day I will find out for sure.

Twinned Twins

Identical twins hold a special place in our imagination. Built from identical DNA, they have exactly the same inherited characteristics. In addition, they are exactly the same age and often attend the same school and have the same friends. They also look identical, so they are often mistaken for each other, further intertwining their lives. (As a child, I had identical twin neighbours who used to play road hockey with us. Since we couldn't tell which was which, we simply called them both "Superstar.")

So, it is not surprising that identical twins share many characteristics. But some people take it a step further and think that twins have some special psychic connection that allows for magical extrasensory communication. They tell striking stories of twins who choose husbands with the same names and professions, or who have very similar habits or preferences, or who experience sympathetic feelings. When New Jersey twins Danielle and Nicole Fisher had babies just 13 minutes apart, they attributed it to "that twin communication."[27] My own neighbour wrote an entertaining book for young adults in which two hockey goalies turn out to be long-lost identical twins, and when one gets injured, the other feels an unexplainable pain.[28] And there is also that old joke about one twin taking a bath and the other twin suddenly getting clean.[29]

Of course, we know by now that such tales involve luck traps. There is *biased observation*, since only striking similarities between twins are remembered—not twin differences, nor similarities between non-twins. And there are *alternate causes* whenever the similarities could be explained by similar upbringings or genetics or appearance. There could also be *false reporting*, since many of these tales are exaggerated. Not to mention *extra-large targets* when counting what is similar—for example, those baby births were 13 minutes apart, not simultaneous. And most importantly, there is a *many tries* luck trap: out of the many millions of identical twins around the world, some of them will have some striking similarities just by dumb luck, signifying nothing.[30]

Despite these various luck traps, the question remains: Can twins actually communicate telepathically? Do they actually have some form of ESP?

In fact, this question has already been tested. Two psychologists conducted an experiment in which one twin was told to try to psychically sense a number, drawing, or picture from the other twin.[31] What they found was interesting. In the first part of the experiment, the first twin was allowed to select the number, drawing, or picture themselves. With this setup, they found that the second twin did indeed guess correctly more often than non-twin pairings or what could be explained by luck alone. This was a striking result. However, an *alternate cause* was at work: the twins' successes could be explained by their having similar habits and preferences, due to identical genetics and similar upbringing. So this experiment provided evidence of what they dubbed "thought-concordance," but it did not demonstrate any actual ESP.

In the second part of their experiment, the choice of number or drawing or picture was instead *assigned* (randomly) to the first twin, who was then told to write it down and concentrate on it. This design eliminated the effects of similar upbringing and so on, since the first twin did not get to select the object. And what did they find? In this case, the twins did no better at guessing the correct object than non-twin pairs or random luck. Once the *alternate cause* was eliminated, there was no evidence of any twin ESP or supernatural communication at all.

Psychic CIA

Despite the failure of many attempts at replication, some people do still believe in certain psychic phenomena. One example is "clairvoyance," the ability to see persons and events that are distant in time or space. And if you really believe in such powers, then this raises the question of whether they could be used for practical purposes. As Robin Williams once put it, "If it's the Psychic Network, why do they need a phone number?"[32] Or, as attributed to comedian Steven Wright, "All those who believe in psychokinesis—raise my hand."[33] All kidding aside, has anyone actually tried to use psychic powers to address specific challenges, such as military intelligence?

Actually, yes. For one, in the 1970s, a US soldier named Jim Channon proposed a "First Earth Battalion" that would replace wars by ethical nondestructive conflict resolution organized along "New Age" principles. Associated with Channon is the martial artist Guy Savelli, who claimed that he could kill goats and hamsters using his thoughts alone, and could kill people just by touching them.[34] Savelli's claims were never verified, but they did provide a focus and title for the book *The Men Who Stare at Goats* by Jon Ronson, later made into a movie of the same name starring George Clooney.

More systematically, the Stargate Project was run by the United States Army under various names from 1978 onwards.[35] It focused on "remote viewing," the supposed ability of some people to sense images and appearances at faraway places. The ultimate goal was to use remote viewing for military intelligence purposes—apparently inspired by fears that the Soviet Union was doing the same. The project commissioned a number of studies by the Science Applications International Corporation (SAIC), and it continued to operate, largely in secret, until 1995. At that point, those working on the project were transferred to the auspices of the CIA, which promptly commissioned a report to determine whether or not the program should continue.[36]

And what did the report say? It considered various experiments the Stargate Project had conducted in an attempt to establish the existence of remote viewing. It noted that in some of those experiments, the subjects had guessed correctly somewhat more often than chance would predict. It then asked two academics for their assessments of psychic powers. One of them, Ray Hyman, concluded charitably that "the case for the existence of anomalous cognition is still shaky, at best," and "the experimental program is too recent and insufficiently evaluated to be sure that flaws and biases have been eliminated." But the other academic, Jessica Utts, came to a very different conclusion. "It is clear to this author," she wrote based on the SAIC experiments, "that anomalous cognition (i.e., remote viewing psychic abilities) is possible and has been demonstrated."

Now, Utts had coauthored many papers with the people who conducted the experiments in question, so she was not neutral on this point. Nevertheless, her remarks are striking. She is an established statistics professor who has written textbooks and published research articles. She knows all about luck traps. I have a special fondness for her since, by coincidence, she once had the role of presenting me with a prestigious academic prize.[37] And it turned out that we have the same birthday, to boot! (Though apparently she was born on a Saturday, not a Friday.)[38] In yet another coincidence, while I was planning this book, she invited me to speak at a large meeting she was organizing, at which she was a delightful and generous host. So, how could I possibly disagree with her about anything?

And yet, Utts's comments about psychic abilities do not convince me. There are indeed some experiments that have shown remote viewing accuracy slightly higher than what would be expected by chance alone. But as we have already seen, these experiments are rife with methodological issues and have failed to be consistently replicated. And, as I have argued, it would take very convincing and consistent evidence to truly establish the existence of psychic phenomena. (Utts herself essentially acknowledged this when she remarked in an article that data demonstrating precognition "would be widely accepted if they pertained to something more mundane.")[39] It still seems to me that, once the luck traps and replication failures are all stripped away, such convincing evidence simply doesn't exist.

I'm not the only one who thinks so. The final CIA report asserted, "It is unclear whether the observed effects can unambiguously be attributed to the paranormal ability of the remote viewers as opposed to characteristics of the judges or of the target or some other characteristic of the methods used. . . . Evidence has not been provided that clearly demonstrates that the *causes* of hits are due to the operation of paranormal phenomena." It concluded that "continued use of remote viewing in intelligence gathering operations is not warranted." At that point, Project Stargate was duly cancelled. Later on, different researchers found other problems with the SAIC experiments, further discrediting them.[40] They have now largely faded from view.

Quite a PEAR

So, did anyone ever replicate the SAIC experiments that had convinced Utts? One group tried. The Princeton Engineering Anomalies Research (PEAR) ran from 1979 to 2007, conducting numerous experiments in an attempt to establish remote viewing and psychokinesis.[41] They claimed, for example, that humans could use their minds to affect the output of random number generators and other random devices. They admitted that "the observed effects were usually quite small, of the order of a few parts in 10,000 on average."[42] This means that the subjects affected only about 0.03 percent of the output, which is virtually none. Nevertheless, they claimed that the overall effect, over a full 28 years of experiments, was still highly statistically significant.

How did the scientific community react to this? Well, one scientist examined the results and claimed that most of the effect was actually the result of a single subject, "Operator 10," believed to be a PEAR staff member, who alone contributed half of the total excess hits.[43] Another scientist asserted that PEAR's "baseline" comparisons were "too good," which skewed the results.[44] A separate analysis indicated that of the 332 experiments performed by PEAR between 1969 and 1987, only 71 gave results consistent with telekinesis, while 261 did not.[45] And most importantly, a later combined attempt by three different laboratories (including, to their credit, PEAR itself) failed to replicate the results, writing that "the primary result of this replication effort was that . . . all sizes of these deviations failed by an order of magnitude . . . to achieve any persuasive level of statistical significance."[46]

PEAR also conducted experiments on remote perception, in which an agent was sent to a random location and asked to observe and record impressions of the details. A faraway "percipient" then tried to sense and describe the location's composition and character and was scored on the accuracy of their description. However, these experiments were widely criticized (including, to her credit, by Utts herself) for "problems with regard to randomization, statistical baselines, application of statistical models, agent coding of descriptor lists, feedback to percipients, sensory cues, and precautions against cheating," rendering their results "meaningless because of defects in the experimental and statistical procedures."[47] These PEAR experiments therefore ended up carrying very little weight.

PEAR was closed permanently in 2007, at which point media reports referred, fairly or not, to its "research that embarrassed university officials and outraged the scientific community."[48] Thus ended the PEAR story once and for all.

Around that time, it was revealed that in 2002 the United Kingdom's Ministry of Defence (MoD) undertook a study to test whether remote viewing (RV) powers could be used to detect hidden objects such as military targets.[49] They spent £18,000 conducting experiments asking recruits to guess the contents of envelopes. They reported that "in the majority of cases the subjects failed to access the target in any degree," and "subjects failed to provide any convincing RV performance." A ministry spokeswoman, asked about the value of the study, admitted, "The study concluded that remote viewing theories had little value to the MoD and was taken no further." Which just about says it all.

CHAPTER 22

Lord of the Luck

I t is hard not to be struck by the incredible richness of the universe. Brilliant stars, lovely lakes, scenic mountains, lush trees, beautiful flowers, delicious food, exotic animals, fresh air, sunshine, and more. And most of all, the astounding complexity of human beings. The feeling of awe is so strong that it immediately leads to other questions: What does this all mean? Where did it come from? Was it all caused just by luck? If not luck, then what?

For many people, the answer is God. It is estimated that over 80 percent of the world's people are religious.[1] Some of them are only mildly religious, following certain practices out of tradition or family pressure or community obligation without completely believing or adhering to religious doctrine. However, many people take their religious beliefs very seriously and use them as a guiding force throughout their lives. And this has lots of positive effects. Religion can provide great comfort, including after the death of a loved one. It inspires lots of people to perform kind, generous, and charitable acts. And just like magic, superstition, astrology, and more, religion also provides a feeling of meaning and purpose and importance for the origin and existence of our incredible world.

By contrast, the scientific perspective is that the universe runs according to certain precise laws and principles. Through the power of such processes as planetary formation, weather systems, continental drift, abiogenesis, microorganism development, animal evolution, and survival of the fittest, over billions of years, the initial big bang gradually gave rise to the planet Earth and all of the amazing things upon it. Indeed, science has managed to come up with precise explanations and convincing evidence for many of these developments, and will presumably continue to do so to an even greater extent in the future.

How does this all fit together? What is the connection between religion and science? Are they compatible, or contradictory? Can we believe in both? Or neither? And most important for this book, how are all of these issues related to luck?

Religion versus Luck?

Religion is a tricky subject to approach. Many people are very passionate and sensitive about it, and can be offended by beliefs that are different from their own. (As the Peanuts comic strip character Linus once wisely noted, "There are three things I've learned never to discuss with people: religion, politics, and The Great Pumpkin.")[2] In many parts of the world, people are oppressed and persecuted because of their religious beliefs. To be honest, I had hoped to avoid discussing this topic. Wars and battles have been fought over religious differences, and I sure don't want to start a new one. However, I have come to realize that I don't really have a choice, since so many people interpret luck in terms of religion.

This point was driven home to me when I taught a freshman seminar class about probability. Unlike most university courses, it had no lectures; instead, there were group discussions about the meaning of probability in our lives. In one class, I mentioned the idea from quantum mechanics that, on the smallest scales of atoms and molecules, physical laws are not clearly deterministic, but rather are based on probabilities and randomness—a concept that has always both fascinated and discomforted me. Could it be, I asked the students, that the scientific laws that govern our universe are really random?

I expected my students to give one of two responses: either "Oh yes, quantum mechanics has shown us the randomness of our world" or "Oh no, surely at the deepest level, our universe is determined unambiguously by clear physical laws." But instead, what they all said, without exception, was, "Well, it depends whether or not you believe in God."

Huh? They had taken the question in a rather different direction than I intended. Their feeling was that if there *is* a God, then God is directing the development of the world, with intention and reason and kindness and caring, and there is no place for randomness to occur. But if there is *no* God, then indeed the universe is just proceeding randomly.

I learned two lessons from our discussion. The first was that, despite my best efforts, my freshman seminar students were never going to understand the idea of quantum mechanics. (Don't feel too bad—we professors experience such disappointments all the time.) The second was that, to many people, religion and luck are sort of opposites—religion is a way for people to *counteract* luck, to argue against it, to take back a feeling of control in their lives. If everything happens for a reason, then it isn't just luck. It has meaning and significance, and maybe even a sort of magic. Which is, as we have seen, what many people prefer.

Religion versus Science?

If religion says that everything happens for a reason according to God's plan, and science essentially says that our universe and existence are all the result of just luck, then are religion and science opposites too? Well, it depends.

Certainly many aspects of religion seem to contradict our modern scientific understanding. For example, the Bible seems to imply that the Earth and all life upon it were created about 6,000 years ago,[3] which directly contradicts the extensive scientific dating of earthly objects and fossil remains much older than that. And the Bible doesn't mention anything about dinosaurs or Neanderthals, which died out long before the books of the Bible were written, but which surely should have been around when "all life" was created. It also reports that the Earth was created before the sun, while ignoring all the other planets and suns and galaxies that we now know are out there. From this perspective, religion and science seem to be quite opposed.

In response to this, many religious people argue that religious documents should not be taken literally, and should not be judged on scientific terms. Rather, religion requires "faith"—an inner belief in the existence of a god that does not require any external evidence to validate. That is, most religious people are content to "know" internally that God exists, without the ability or desire to *demonstrate* this fact. This approach moves religion outside of the domain of science or logical analysis. As Alan Arkin's judge character declares towards the end of the movie *Thirteen Conversations about One Thing* (to explain why a defendant in a trial cannot be convicted based on religion), "Faith is the antithesis of proof."[4] In other words, religious faith is completely separate from issues of evidence and logic and demonstration—and, I guess, of science itself.

In addition, science is weakest where religion is strongest. Namely, science omits any discussion of *why*. Those scientific laws and principles proceed without regard to questions of reason, justice, morality, justification, or right and wrong. Things don't happen for a reason; they just happen. As a physicist character in the movie *September* once said, "It doesn't matter one way or the other . . . it's all random . . . resonating aimlessly out of nothing, and eventually vanishing forever . . . all space, all time, just temporary convulsion . . . haphazard, morally neutral and unimaginably violent."[5]

By contrast, religion can provide great comfort. Two of my friends from university science classes ended up marrying each other, buying a house, and becoming the proud parents of two fine sons. So, did they live happily ever after? Unfortunately not. By incredibly bad luck, the father got a rare form of leukemia. He underwent various experimental medical treatments (I even drove him to a few of them), but in the end they all failed and he passed away far too soon. The family had a religious funeral service at a local Christian church, where various speakers assured us that the death had happened for a reason and that the father had gone off to a better place. Despite my own lack of religion, I was glad the family took comfort in their belief. Maybe it didn't matter if this belief was "true," or if I shared it; it was more important that this belief helped my friends get through a time of great sadness.

So perhaps that is the resolution to the relationship of religion to science. Perhaps religion is based purely on an inner belief that God exists, without any evidence or proof or demonstration about ultimate "truth." Perhaps that faith gives people comfort and meaning and wise guidance for their life choices, rather than precise, logical insights. Perhaps religion focuses on the big "why" questions of meaning and purpose and so on, while science focuses on the specific "how" questions about the rules and principles that govern the day-to-day progression of the universe. Perhaps science and religion aren't in contradiction after all. And regarding luck, perhaps science shows us the fundamental principles by which luck occurs, while religion provides us with the hidden meaning behind the apparent luck that we observe.

Perhaps. But not necessarily. Some religious people and groups aim to convert others to their own belief system, or convince non-believers that they should become religious, or argue that rules and laws should be governed by religious doctrine. In that context, it seems relevant to consider, from various perspectives, the question of actual *evidence* for the existence of God, and the corresponding meaning and purpose of luck. So, what can we say about that?

Atheists' Nightmare?

To begin with, some people claim very direct evidence for the existence of God. To take one example, the evangelist Ray Comfort argued on television that bananas—that's right, the long, thin, yellow fruit—are an "atheist's nightmare."[6] Why? Because bananas are so conveniently designed, notably by being so easy to peel due to their protruding little "tab" on top, that they could only have arisen from God's hand. This, he insists, provides conclusive proof, once and for all, that God exists. Well, does it?

My first reaction was that the video's use of the phrase "atheist's nightmare" is odd. Atheists do not live in fear of evidence of the existence of God; they simply do not believe that any clear evidence exists. If clearly convincing evidence did arise, then most atheists would accept and welcome it, rather than consider it to be a nightmare. This illustrates just how wide a gulf exists between evangelists and atheists.

But let's leave that aside. The real question is: Do bananas prove the existence of God? Or is this argument actually, well, bananas?

The argument has certain immediate limitations. For one thing, bananas from earlier centuries were differently shaped and harder to peel. The nice, convenient bananas we eat today are the product of not just nature, but also the *alternate cause* of years of agricultural engineering. (Of course, one could argue that this agricultural engineering was carried out by humans who were in turn created by God, who therefore still deserves the credit, but that seems like quite a stretch and rather different from what the original claim intended.)

Also, bananas can still be a bit tricky to open (especially if they are a bit underripe or overripe), don't always hold their shape, and often leave little stringy bits that have to be removed separately. I'm not complaining—I love bananas—but claims that bananas are truly so easy to peel might involve a bit of exaggeration or *false reporting*. Put another way, if an all-powerful being had truly designed bananas specifically to be easy to peel, then they could have done a better job.

In addition, the convenience of eating bananas is consistent with the theory of evolution. Indeed, traditional bananas (if not the latest engineered variety) contain seeds.[7] So the easier bananas are to eat, the more that animals and humans will eat them and spread their seeds around to grow more banana trees, and the more the plant will flourish. It actually makes perfect sense from a purely scientific perspective, thus providing a different kind of *alternate cause* for an easy-to-peel fruit.

But these are all just minor quibbles. The weakest point in Comfort's argument comes from other fruit besides bananas! After all, if God took the trouble to design the banana to be easy to peel, then surely he should have made a similar effort with, say, oranges. But oranges do not have any helpful "tab" on top, and indeed many of them are quite difficult to peel without a sharp knife. So this is a form of *shotgun effect*: out of so many different kinds of fruit, just one has a helpful tab for easy peeling. Therefore the argument that the tab is a divine innovation is less convincing than it would be if *all* fruits offered similar convenience. Once again, it all comes down to luck traps.

Evil Evidence

Of course, most people do not believe in God because of bananas, or any other single specific item. Rather, much of religious belief comes from observing all the wonderful, amazing things in our world and thinking that it must be the result of more than luck. Not to mention human beings, who are so mind-bogglingly complex that it seems inconceivable that they would arise just by chance. Surely there is a reason, a purpose, an intention to it all. Surely this is not just luck.

Surely this is evidence of a generous God who created all these wonderful things.

I find this argument somewhat compelling. It is indeed astounding to consider all of the wonderful and beautiful and complicated and sophisticated things that exist in this world. Could they really be the result of just *dumb luck*, of the randomness of scientific forces like planetary formation, ecological drift, animal evolution, and survival of the fittest? It is easy to see why many people "feel" the presence of a purpose to it all, an intention in the design, a force that created it all.

However, this argument has one flaw. Namely, there are also many very *bad* things in this world, including untold human suffering from such natural phenomena as earthquakes, floods, diseases, and famine. So many unnecessary deaths, so much pain and misery. If God were truly all-powerful and infinitely kind, then surely everything on Earth should be good, and nothing at all should be bad, which just isn't the case. (This is related to the philosophers' logical argument called "the problem of evil.")

In response, some argue that God gave us free will, which "explains" the tragedies caused by humans—things like wars and murders and oppression. Okay, fine. But what about the tragedies caused not by humans but by forces of nature? How, in the presence of an omniscient and omnipotent and omnibenevolent God, can they be explained? Some might claim they are the work of the devil, or God's way of testing us, or that God works in strange and mysterious ways that we cannot possibly understand. But none of those responses seems very convincing. On the contrary, they suffer from a *biased observation* luck trap where God is given credit for all the good but no blame for all the bad.

The comic actor (and avowed atheist) Stephen Fry argued similarly. When asked on Irish television what he would say to God if he ever met him, he replied that he would complain about such evils as bone cancer in children and eye-destroying insects, implying "a capricious, mean-minded, stupid God" whose actions were "simply not acceptable."[8] (Upon hearing of these remarks, the Irish police actually launched a criminal investigation of Fry, for suspected violation of that country's blasphemy laws, though they later called off the investigation. I sure hope that, upon reading this chapter, the police don't start investigating me too.)

I understand Fry's anger, but I do not share it. I think the world is a wonderful and fantastic place. Despite the terrible sufferings and injustice, overall I can't complain (though perhaps this is simply because I am not among the most oppressed), and I think that this world, despite its many flaws, is much preferable to no world at all. I feel very grateful and *lucky* for the life I have been allowed to live. On the other hand, when it comes to the question of evidence, it seems to me that a world with some good, but also some bad, is actually more consistent with the effects of luck and randomness than with the existence of an all-good, all-powerful creator who is constantly watching over us.

Moral Morass

Another argument for religion is that God is necessary for morality—that without God we cannot possibly figure out right from wrong. For example, the talk show host Steve Harvey (best known for announcing the wrong winner of the Miss America contest in 2015[9]) boldly stated in an interview, "You sittin' up here talking to a dude, and he tells you he's an atheist? You need to pack it up and go home! You know, you talkin' to a person who don't believe in God? What's his moral barometer? Where's it at? It's nowhere!"[10]

By one estimate, there are nearly half a billion atheists around the world.[11] This includes most leading scientists: recent surveys suggest that only 7 percent of members of the US National Academy of Sciences believe in a personal God,[12] while 64 percent of Fellows of the UK Royal Society strongly disagree that God exists (compared to just 5 percent who strongly agree).[13] The half billion also includes lots of celebrities, including Kevin Bacon, Jodie Foster, John Malkovich, Björk, Diane Keaton, and Daniel Radcliffe.[14] Mr. Harvey is thus declaring that hundreds of millions of people he has never met are all completely amoral, to the point where no one should even agree to talk to them. This is actually quite insulting, and I hope and believe that most religious people would not take such a hard and uncompromising line. But setting aside the extremity of the remarks, we can ask: Is it really so? Is it really true that a belief in God is required in order to hold moral principles?

Certainly, many religious people are very moral. Numerous religious-based charities in every corner of the world devote significant money, time, and effort to helping feed the hungry, house the vulnerable, and teach the disadvantaged, and I salute them all. And many brave religious people, such as the liberation theologists in Latin America and elsewhere, have endured great risks and tremendous sacrifices to help relieve the suffering of poor and oppressed peoples.

But not all religious people are so moral. In fact, religion has sometimes inspired very evil acts, from crusaders in the Middle Ages who slaughtered multitudes of Muslims in the name of Christianity, to European settlers of the Americas who oppressed and largely destroyed indigenous cultures in the name of Christian "civilization," to modern-day suicide bombers who kill thousands of innocent civilians in the name of Islam. Sexual abuse of children by Catholic priests has been widely documented[15] (in fact, the day after I first wrote this sentence, it was announced that the third-highest-ranking official in the Vatican, Cardinal George Pell, had been charged with multiple sexual assaults[16]). The violent, racist Ku Klux Klan movement claimed to uphold Christian morality. Brutal Hindu–Muslim fighting ended an estimated half million lives during the 1947 partition of India alone.[17] Many thousands have been killed, and many more displaced, in continuing hostilities between Jews and Muslims in the Middle East's never-ending Arab–Israeli conflict.

The Northern Ireland civil war "Troubles" between Protestant and Catholic religious forces took more than 3,500 lives.[18] Religion is often used to justify the oppression of groups based on gender, race, or sexual orientation. And the perpetrator of the senseless Sandy Hook shooting of 20 innocent schoolchildren was a parishioner at his local Catholic church.[19] In fact, in a 1997 study, 7 percent of university students said they would kill if God told them to.[20] These and other examples illustrate that, at the very least, religion does not always guarantee moral behaviour.[21]

And many non-religious people are very moral. For example, Brad Pitt and Angelina Jolie do not believe in God at all,[22] but they have given many millions of dollars to charitable causes (both before and after their divorce).[23] Similarly, Warren Buffett says he does not know if there is a God,[24] but that has not stopped him from donating billions of dollars to charities.[25] In fact, one analysis concluded that if donations to congregations and religious organizations are excluded, then non-religious people actually donate more to charities than religious people do.[26] And a recent study of over a thousand children in six countries concluded that non-religious children are actually *more* likely than religious ones to share their possessions.[27] These examples are not the final word, but they do indicate that many non-religious people behave in ways that are very generous and moral.

Some religious morality claims involve severe luck traps. For example, Tony Perkins is the president of the Family Research Council,[28] a conservative Christian organization that asserts that "homosexual conduct is harmful to the persons who engage in it and to society at large."[29] He argued in a 2015 interview that Hurricane Joaquin, which caused great damage to several Carribean islands and killed 33 people on an American cargo ship, wasn't bad luck but was actually a sign.[30] Since the hurricane was originally headed towards "the nation's capital" of Washington, D.C., it represented "God sending us a message" in response to humanity's "act of desecration" in legalizing gay marriage, and thus "turning the holy vessel of marriage . . . against its purposes." Then, in 2016, his own home in southern Louisiana was destroyed by massive flooding, forcing him to live in a camper for six months.[31] Was this another message from God in response to desecration? No, this flooding was an "incredible, encouraging spiritual exercise." In his *biased observation*, what happened to others was God's justifiable wrath, but what happened to him was a positive development.

My only point here is that the relationship between religion and morality is not so clear-cut. In the end, both good and evil actions can come from both religious and non-religious people. Religion may often lead to moral behaviour, but it is neither required nor a guarantee. As Albert Einstein himself once put it, "A man's ethical behavior should be based effectually on sympathy, education, and social ties and needs; no religious basis is necessary. Man would indeed be in a poor way if he had to be restrained by fear of punishment and hopes of reward after death."[32] If so, then true morality is separate from religion and does not provide any clear evidence either for or against God's existence.

And what about myself? Well, I am not religious, but I certainly try to act morally. I am honest, law-abiding, and hard-working, and I take my responsibilities very seriously. I make a concerted effort to treat other people with respect and kindness, and to help them out when I can, not out of religious obligation but out of a genuine belief in the value of humanity (this approach is sometimes called secular humanism). I give significant amounts of money each year to charities that provide food and housing for the less fortunate. I think hard and often about how to assist my family and friends and neighbours and colleagues. I often find myself trying to decide what is the most appropriate or fair or generous action to take in a given circumstance. And if I ever accidentally offend or inconvenience someone, then I feel genuine regret.

Am I perfect? Not at all. Have I made great sacrifices, like liberation theologists? Certainly not. Is my morality anything special? No, not in the least. But I do try, in my own limited and feeble way, to be a good person and do the right thing. Despite my flaws, I am fairly certain that if a religious person got to know me well, they would decide that I am indeed moral. I hope that, despite my lack of religion, they would find me to be a good and kind person that they could get to know and like and respect.

Of course, Mr. Harvey would never find out any of this. He would refuse to talk to me at all, simply because I do not share his beliefs. Is that his best example of moral behaviour? I do not know Harvey at all, so I will not judge him. But perhaps he shouldn't judge me either.

Lucky Earth?

Even if there isn't any clear evidence about the existence of God or some external factor causing the origin of life on Earth, it is still hard for many people to believe that our wonderful planet and species arose just by luck and scientific processes alone. Can our understanding of luck traps be of any help?

Well, there's no *extra-large target* here. No one can say it was easy to create the detailed beauty and complexity and sophistication of our planet's mountains, lakes, plants, animals, and, especially, humans. No one can say that any other planet would have done just as well. No one can say that the details are flexible or unimportant. For us to be here, every aspect had to be just right.

Also, there is no *false reporting* or *placebo effect* involved. We humans are most definitely here, living on this planet, eating and breathing and moving and thinking and thriving. No doubt about that, and no debate there.

There is, however, a form of *hidden help*. Namely, the theory of evolution provides a subtle and powerful mechanism by which new types of life are created. First, some genes are randomly mutated from one generation to the next. Then, those mutated offspring either die out (if the mutation is not helpful) or flourish (if it is). Over the course of millions and millions of years, some species gradually branch off and give rise to new and different species that are better suited to their environments. There is now overwhelming scientific evidence for evolution,[33] and it was clearly essential in bringing our planet to its current state of intelligent life.

Another luck trap is even more significant: the *shotgun effect*. What are the different pellets of lead here? Why, all the different planets where life *could* have evolved. Carl Sagan famously lectured about the billions of galaxies, each with billions of stars, that make up our universe.[34] Since his time, the estimated number of stars has only increased, and in fact might even be infinite.[35] We do not know how many planets orbit all those stars, but since thousands of planets have already been detected,[36] there are almost certainly billions of billions—or perhaps even an infinite number—of planets too. So, when considering the fact that intelligent life evolved on our planet, we need to ask: On one planet *out of how many*? With so many billions of available planets out there, it becomes at least somewhat less surprising that one planet developed life.

These scientific explanations and luck-trap considerations are sufficient to convince my *brain* that our incredible planet and amazing species could indeed develop through a combination of evolution, randomness, and genuine luck. I believe that we can act morally, and find meaning and joy in our lives, from a direct secular humanist belief in the value of humanity, and through such genuine human emotions as compassion, love, respect, and ambition, all without the need for religion. And that is why I personally do not believe in a God, though I certainly respect those who do.

But my *heart*? Well, my heart still can't really believe that we're here.

CHAPTER 23

Lucky Reflections

This book began with the question of whether I believe in luck. I said that it depends what you mean. And after writing a whole book on the subject, I guess I have to stick with that same answer.

I can say that I recognize what an important role luck plays in our lives, from who meets whom, to who succeeds and who fails, to who lives and who dies. Some of this luck is beyond our control—just dumb luck—while some of it can be influenced by our choices and planning and preparation. And, as the luck version of the Serenity Prayer says, it is useful to know the difference. That isn't easy, but by carefully examining the evidence and avoiding the multitude of luck traps that we have seen, we can try to sort out which factors truly influence events and which do not.

Joy without Magic

In the movie *Heaven Can Wait*, the Warren Beatty character is removed from his body by a well-meaning but misguided "escort" as he is about to be hit by a car. It turns out that he actually would have avoided the car—the escort was mistaken, as confirmed when they later look up the character's actual death date and find that it is many years in the future. This tragic error is explained using the phrase "probability and outcome." That is, the escort was summoned because there was a certain probability that the character would die, but then, in the outcome, the character actually survived.[1]

Now, the movie's logic is a bit dubious. How could escorts know in advance when someone will probably die, but not when they will actually die? And, if they can later "look up" the latter information, then why did they make their error in the first place? Nevertheless, the phrase they used stuck with me: probability and outcome. Chance and result. Randomness and reality. If my lifelong work—the mathematical theory of probability—is about the chances, then perhaps luck is about the results. Perhaps luck is what actually happens, after all the probabilities have been set in place. Perhaps people are willing to let probabilities be determined by the genuine rules and science and facts and figures of the world we inhabit, but want to control the actual outcomes nevertheless.

Luck and magical influences have a great attraction, in fiction and beyond. Indeed, we might all *wish* that our world had genuine magic, supernatural abilities, and built-in meaning. It would be exciting, and different, and give us special reason for living. That's why we write our fiction that way.

But in the end, it is also satisfying, in a different and perhaps deeper way, to understand the difference between fiction and reality.

Indeed, imagine if reading from *Macbeth* really could cause bad things to happen. Or that praying could make a diamond jump from the road to a shirt fold. Or that dice really did conspire to choose winners and losers. Or that people really did magically receive signals of external information while sleeping. If any one of these things were true, it would be by far the most significant scientific development in human history. Atom bombs and transistor radios would pale by comparison. Computers and microwave ovens would be afterthoughts. Worldwide headlines about the Higgs boson or gravitational waves would be trivialities. Everything that scientists thought they ever knew about our world would be completely upended and revised. This is why tales of magical forceful luck are so intriguing, but also why we should be cautious about accepting such claims, and why the standards of proof have to be so high.

Yes, the world is full of mysteries about human consciousness, nature's complexity, emotions, and more. But those strange and surprising stories about unusual coincidences, good luck charms, magical influences, and divine intervention? I do believe that the evidence shows that they can all—yes, *all*—be explained by a combination of luck traps, selective observation, scientific causes, and human thinking. Not a very magical perspective, but nevertheless an accurate summary of the luck in our world. Despite this realization, I am confident that we can accept our world as it is and still find joy, happiness, and success. How? Through *non*-mystical forces such as friendship, love, hard work, practice, education, experiment, and logical thinking.

So I will continue to enjoy the power of the witches' prophecies in my favourite Shakespeare play, *Macbeth*. But I will also remain confident that in the real world, prophecies have no magical power, superstitions do not control us, and it is okay to quote from a play if the time is right.

A Fortuitous Flick

After submitting the first complete draft of this book to my editor, that very evening I searched for a diversion on Netflix. I eventually settled on a recent British movie called *The Hippopotamus*,[2] which I hadn't previously heard of but which sounded interesting enough. Little did I know that, *as luck would have it*, this flick would end up nicely encapsulating my entire book.

In the movie, a drunken old former poet is contacted by his goddaughter with some astonishing news. She explains that she had been diagnosed with terminal cancer and had just three months to live.

Then, her cousin—the poet's godson—used his special powers of touch to miraculously cure her of her disease. She is intrigued and manages to hire the poet to go and visit the godson's estate, to clandestinely discover the true secret behind his powers.

After a slow start, the poet eventually witnesses the godson performing various other miraculous cures. The godson's mother once had terrible asthma and nearly died, but the godson's magic touch vanquished the asthma once and for all. A prize horse is sick and about to be put down, but astonishingly recovers the day after the godson slips into the barn. A house guest's serious angina pain disappears, and he happily tosses his pills into the garbage. Next up, the godson is asked to take a visiting "plain" French teenager and make her beautiful. The poet, initially skeptical about the existence of these special powers ("I don't believe in miracles!"), starts to become convinced.

As the story continues, we learn more about the family history—apparently, the godson's grandfather had similar healing powers. There is also some further family mystery, and suggestions that perhaps the godson's straitlaced older brother is secretly assisting him. The big question, of course, is how the godson's powers actually work, and who he might cure next. I was suitably absorbed by this plot, and saw it as another nice illustration of how in fiction (as opposed to real life), magic and supernatural powers have their place and often add excitement and mystery. So far, so good.

But then the movie ends with an unexpected twist. (Spoiler alert!) The poet realizes that the prize horse wasn't actually sick, but merely hungover after accidentally drinking a lot of whiskey (long story), so the horse's recovery wasn't miraculous after all. From there, the poet discovers that the mother's asthma attacks hadn't really ceased; she just said so to please her son. And the house guest's angina pain returns too—he had merely convinced himself that it was gone. Oh, and the goddaughter whose cured cancer had launched his investigation?

Sure enough, she was merely in remission; her cancer soon returns and becomes fatal.

In short, the godson's special powers were really nothing more than a combination of *false reporting*, *alternate cause*, and *placebo effect*, with nothing supernatural about them. The godson believed in his special powers, and so did all those around him, but they were all wrong. To my surprise, the movie ultimately wasn't an example of magic in fiction, but rather a cautionary tale about the perils of luck traps!

After all that, I suppose I shouldn't have been too surprised to discover that the movie was actually based on a novel by our old atheist actor friend Stephen Fry.[3] Like the film's poet, I guess he doesn't believe in miracles either.

A Painful Goodbye

Around the time I was first planning this book, my family received some very unlucky news. My dear mother, otherwise very healthy and active, was diagnosed with pancreatic cancer. A quick Google search revealed my worst fears: pancreatic cancer is among the most aggressive types, with a survival rate of only about 14 percent.[4] Although my mom's cancer was discovered relatively early, and although she soon had surgery (a Whipple procedure) to remove the tumour, and although that was quickly followed by six months of intensive chemotherapy, nevertheless the cancer soon returned. My mother gradually got weaker, and eventually was moved to a hospital palliative care unit.

On Saturday, August 19, 2017, at 5:40 a.m., I received a phone call. The nurse on duty noticed that my mother's breathing was becoming irregular and thought she didn't have long to live. So my wife and I jumped in the car and drove to the hospital to be with her in her final hours, while frantically phoning other relatives to spread the word. My mom took her final breath about 80 minutes later—just a few weeks before I finished the first draft of this book—with me and other loved ones sitting by her side.[5]

What am I to make of this? Was there a "reason" my mom got cancer? Was she being "punished" for past deeds? Was it her "destiny"? Was it part of a master "plan" that we can only guess at? No, I don't think so. I think it was just really, really bad luck.

Interestingly, when I got that phone call at 5:40 a.m., I wasn't asleep. I had woken up about 15 minutes earlier with a bit of stomach discomfort. Was this my mother's way of "communicating" with me just before she passed? Was this my body "sensing" her impending doom and reacting sympathetically? Did this illustrate a secret, hidden, special "connection" between me and her? No, I don't think so. I think I was just worried about my mom's condition, which caused my stomach to tighten up.

Just after my mom died, while we were all sitting beside the hospital bed, an aunt remarked that it was nice to think my mom might be with her friends who had also passed, in some sort of afterlife. I'm sure my aunt was trying to be helpful, but I took no comfort from her words. I saw no evidence that my mother had gone anywhere other than simply lying dead in her hospital bed. If she were truly being taken to a better place, I thought, then surely they could have taken her *before* she suffered through all the cancer and treatment discomfort.

Would it have been easier for me if I believed in some sort of destiny, some meaning for her death? Or if I took my upset stomach as a sign from the great beyond? Or if I believed my mom had gone on to an eternal life of bliss in heaven above? Perhaps. It's hard to say.

But I do know that, without resorting to any supernatural explanations, I was able to appreciate my mother's life well lived, relive fond memories of our happy times together, and accept death as an inevitable part of life. I could find my own sense of meaning and joy in everything we had shared together, everything she had done for me, and all the ways she had made my life better. I could celebrate her life and mourn her death with pride and resolve, without sacrificing any of my scientific perspective. And *luckily*, she would have wanted no less.

Dour Luck?

Whenever I try to sort out the world logically in terms of what causes what, and what doesn't cause anything, there are inevitably some people who complain that I am making the world less fun. "Aren't you being a killjoy?" they ask.

They hope that wishing hard will help them win the lottery jackpot, while I use harsh statistics to tell them they're more likely to die on the way to the store. They imagine that "fate" will help them find their perfect life partner, while I coldly compute the probability of running into someone unexpectedly. They fantasize about magic forces controlling their destiny, while I insist that our lives are governed by unfeeling physical laws of nature. "Is that all there is?" they ask.

There are many reasons why people believe in some form of supernatural, forceful luck. Perhaps they want events to have meaning, like in so much fiction, instead of being just random. Perhaps they take comfort in the thought of a purpose or grander scheme for the misfortune they encounter. And, perhaps they have *biased observation*, only noticing and remembering those events that confirm their beliefs, while ignoring events that don't. Also, they probably interpret good and bad luck very broadly, thus creating an *extra-large target* in which they can always find *something* to justify their beliefs.

I certainly understand why people hold these beliefs. However, I do not share them. Why? Because there is no convincing *evidence* for them. Virtually all of the observations that suggest supernatural phenomena can be explained away by the various luck traps that we have explored. And if there is no clear evidence, then I cannot find it in me to believe in these outlandish theories.

I know that many people find this disappointing. But I don't think they should. I think that the world we all inhabit, even without any supernatural phenomena, is already full of joy and wonder. There are billions of people out there whom you might meet and get to know and share with and care about. There is an unending list of fascinating facts and subjects to learn and ponder. There are books, plays, and movies of every sort and taste. You can sing and dance, laugh and cry, love and hate, support your family and be kind to strangers, all within the scientific parameters of the universe as it actually is. You can imagine and dream, and enjoy made-up stories about magic and movies full of fantasies, knowing that they are different from real life.

I myself have performed music and comedy, teared up at movies, fallen in love, made lifelong friends, cared passionately about work and hobbies, and been gripped by literature—from the prophetic witches in Shakespeare's *Macbeth* to the magical dreams of W.P. Kinsella's *Shoeless Joe*—without once feeling that the scientific laws of the universe prevented my fun and joy. Not bad for a guy who was born on Friday the Thirteenth.

I firmly believe that the world can and should be considered the way it actually works, without appeals to supernatural forces. I think this leads to more logical thinking and better decisions, while still allowing for incredible and wondrous experiences on this vast planet we all inhabit. And if this book has made you think just a little bit differently about the luck and randomness and uncertainty that we all experience every day of our lives, then I will consider myself to be *very lucky indeed.*

Acknowledgements

I am pleased and honoured that the team from my first book is still with me, including my editor, Jim Gifford; my agent, Beverley Slopen; my wife, Margaret Fulford; the publishing team at HarperCollins Canada; and all of my wonderful family and friends and colleagues. Thanks for everything you have done for me. I feel very lucky to have all of you in my corner.

About the Author

Jeffrey S. Rosenthal is a professor of statistics at the University of Toronto. Born in Scarborough, Ontario, Canada, he received his B.Sc. in mathematics, physics, and computer science from the University of Toronto at the age of 20; his Ph.D. in mathematics from Harvard University at the age of 24; and tenure at the age of 29. He was awarded the 2006 CRM-SSC Prize, the 2013 SSC Gold Medal, Fellowship of the Royal Society of Canada and of the Institute of Mathematical Statistics, and also the 2007 COPSS Presidents' Award, the most prestigious honour bestowed by the Committee of Presidents of Statistical Societies. He has received teaching awards at both Harvard and Toronto.

Rosenthal's first book for the general public, *Struck by Lightning: The Curious World of Probabilities*, was a bestseller in Canada, and was published in 14 countries and ten languages. Rosenthal has also published two textbooks about probability theory, and over 100 papers in refereed research journals, the majority related to Markov chain Monte Carlo (MCMC) randomized computer algorithms.

Rosenthal has frequently been interviewed on radio and television, and has given numerous presentations to general audiences, discussing various aspects of probability, statistics, randomness, and mathematical reasoning in everyday life. He has also worked as a computer game programmer, musician, and improvisational comedy performer. His website is probability.ca, and on Twitter he is @ProbabilityProf.

Despite being born on Friday the Thirteenth, Rosenthal has been very lucky.

Glossary

Here is a brief glossary of some of the terms used in this book.

General Terms

dumb luck or **random luck.** Luck that occurs completely by chance, with no particular meaning or cause.

forceful luck. Luck that is the result of special, magical, ethereal forces that control and influence events.

just luck. A result that can be explained entirely by luck traps, proving nothing.

luck. Outcomes (good or bad) that are important to us, but are outside of our direct control; either random luck or forceful luck.

luck trap. A situation that can trick us into misinterpreting evidence and drawing false conclusions.

p-value. The probability that an outcome would have occurred in the absence of any specific cause, just by luck alone.

statistically significant. When an outcome is very unlikely to have occurred by luck alone, usually meaning that the p-value is less than 5 percent (or sometimes 1 percent).

Specific Luck Traps

alternate cause. When an effect is the result of a different phenomenon than it initially seems to be.

biased observation. When certain evidence is taken into account while other evidence is missed or ignored.

different meaning. When an effect actually demonstrates something quite different from what it initially appears to demonstrate.

extra-large target. When an effect is actually much easier to achieve than it initially appears.

facts that go together. When multiple effects are all closely related, and hence don't provide as much additional evidence as they seem to.

false reporting. When the facts are different from what is claimed.

hidden help. When some other non-obvious factor makes an effect much more likely than it would otherwise be.

lucky shot. When an event occurs purely by random luck alone.

many people. When lots of different people have separately attempted to achieve the same effect.

many tries. When there were lots of separate opportunities to achieve the same or similar effect.

out of how many. The consideration of the number of different possible combinations that each could have achieved a similar effect.

placebo effect. When observations are influenced by psychological factors.

shotgun effect. When there were lots of different ways to potentially achieve a similar effect.

to multiply or not to multiply. The issue of whether different facts tend to go together, so the chance they will all occur at once is higher than simply the multiplication of all of their individual probabilities.

Notes and Sources

Here are some further details and sources regarding some of the content of this book. Wherever possible, I provide web links for online viewing. Of course, due to the dynamic nature of the web, some of these links might be out of date by the time you visit them. Also, some of the web addresses may be awkward to type into your computer; to help with this, a version of these notes with clickable web links is available at http://www.probability.ca/kow.

Chapter 1: Do You Believe in Luck?
1. The interview was for the CBC Radio program *Fresh Air*, with host Mary Ito, on March 16, 2014 (the day before St. Patrick's Day).
2. Dictionaries don't help much either. For example, Merriam-Webster (https://www.merriam-webster.com/dictionary/luck) first defines *luck* as "a force that brings good fortune or adversity"—in other words, forceful luck. But they quickly follow with a second definition: "the events or circumstances that operate for or against an individual," which sounds like dumb luck with no special force or meaning.

Chapter 2: Lucky Tales
1. Burns's 1785 poem "To a Mouse, on Turning Her Up in Her Nest with the Plough" can be found at: http://www.robertburns.org/works/75.shtml.
2. See e.g. Patrick Bernauw, "The Curse of *Macbeth*," *Unexplained Mysteries*, August 10, 2009, https://www.unexplained-mysteries.com/column.php?id=160421.

3. The program was hosted by Dini Petty on radio station 1050 CHUM in Toronto. My interview was recorded on September 13, 2005, and was broadcast on October 14, 2005, and again on October 16, 2005.

4. One example of such a craps "school" is this class offered by Dice Coach: https://www.dicecoach.com/settingclass.asp.

5. See e.g. Mathew Jedeikin, "The Creepy Reason Why a Rabbit's Foot Is So Lucky," *OMG Facts*, September 21, 2016, https://omgfacts.com/the-creepy-reason-why-a-rabbits-foot-is-so-lucky.

6. See e.g. "Four Leaf Clover," GoodLuckSymbols.com, https://goodlucksymbols.com/four-leaf-clover/ and Acik Mardhiyanti, "The Meaning of Four-Leaf Clover," *My Wonderful Journey*, April 9, 2016, http://acikmdy-journey.blogspot.ca/2016/04/the-meaning-of-four-leaf-clover.html.

7. See e.g. "Horseshoe Superstition," *Psychic Library*, http://psychiclibrary.com/beyondBooks/horseshoe-superstition/; W.J. Rayment, "Luck and Horseshoes," InDepthInfo.com, http://www.indepthinfo.com/horseshoes/luck.htm; and "Good Luck Horseshoe," GoodLuckSymbols.com, https://goodlucksymbols.com/good-luck-horseshoe/.

8. See e.g. "Ladder Superstition," *Psychic Library*, http://psychiclibrary.com/beyond-Books/ladder-superstition/; Debra Ronca, "Why Is Walking under a Ladder Supposed to Be Unlucky?," *How Stuff Works*, August 6, 2015, https://people.howstuffworks.com/why-is-walking-under-ladder-unlucky.htm; and Chris Welsh, "Walking Under a Ladder," *Timeless Myths*, updated April 24, 2018, http://www.timelessmyths.co.uk/walking-under-a-ladder.html.

9. See e.g. Evan Andrews, "Why do people knock on wood for luck?", *History*, August 29, 2016, https://www.history.com/news/ask-history/why-do-people-knock-on-wood-for-luck.

10. See e.g. Hannah Keyser, "Why Do We Cross Our Fingers For Good Luck?" *Mental Floss*, March 21, 2014, http://mentalfloss.com/article/55702/why-do-we-cross-our-fingers-good-luck.

11. See e.g. Debra Ronca, "Why Do People Throw Salt over Their Shoulders?," *How Stuff Works*, August 6, 2015, https://people.howstuffworks.com/why-do-people-throw-salt-over-shoulders.htm. The spilled salt at the Last Supper was apparently invented by Leonardo da Vinci when he painted his famous portrait, leading to this widespread superstition, even though da Vinci himself was not superstitious—see Ross King, *Leonardo and the Last Supper* (London: Bloomsbury, 2012), 234–35.

12. The belief that birds have special divining powers, which can be tapped by stroking the furcula (wishbone), dates back to Etruscans in around 700 BC. See Debra Ronca, "Why Are Wishbones Supposed to Be Lucky?," *How Stuff Works*, August 18, 2015, https://people.howstuffworks.com/wishbones-lucky.htm and Bryan Adams, "The Wishbone Tradition—The Lucky Break," American Academy of Estate Planning Attorneys, November 22, 2010, https://www.aaepa.com/2010/11/wishbone-tradition-the-lucky-break/.

13. According to Rodika Tchi ("Jade Meaning—Ancient Strength and Serenity," *The Spruce*, December 6, 2017, https://www.thespruce.com/jade-meaning-ancient -strength-and-serenity-1274373), jade is "a good luck feng shui stone" that is "employed for various purposes—from creating wealth to attracting more friends."

14. See "Black Cats—Lucky or Unlucky?," International Cat Care, https://icatcare.org/ black-cat-week-unlucky-or-lucky; "Those Lucky Black Cats," *Full Circle*, August 27, 2007, http://fullcirclenews.blogspot.ca/2007/08/black-cats.html; and Becky Pemberton, "Supurrstitions: Is a Black Cat Crossing Your Path Bad Luck or Good Luck? The Superstition Explained," *Sun* (London), October 31, 2017, https://www .thesun.co.uk/fabulous/4676046/black-cat-crossing-path-good-luck-bad-luck- superstition-explained/.

15. The Serenity Prayer is attributed to Reinhold Niebuhr in 1932. See Fred Shapiro, "Who Wrote the Serenity Prayer?," *Chronicle of Higher Education*, April 28, 2014, http://www.chronicle.com/article/Who-Wrote-the-Serenity-Prayer-/146159/.

Chapter 3: The Power of Luck

1. Parker was on Waikiki Beach, just southeast of Honolulu.

2. See e.g. Brian R. Ballou, "On a Beach, Brotherhood," *Boston Globe*, April 28, 2011, http://archive.boston.com/news/local/massachusetts/articles/2011/04/28/ twist_of_fate_brings_half_brothers_together_in_hawaii/; "'Do You Mind Taking Our Picture?': Long-Lost Brothers Reunited by Photo-op in Hawaii," *Daily Mail* (London), April 30, 2011, http://www.dailymail.co.uk/news/article-1382303/ Do-mind-taking-picture--Long-lost-brothers-reunited-photo-op-Hawaii.html; Lynne Klaft, "Chance Unites Brothers," reproduced on the WutangCorp.com forum at http://www.wutang-corp.com/forum/showthread.php?t=108639; and Adam Hunter, "Inexplicable Coincidence on a Hawaiian Beach," *Guideposts*, May 5, 2011, https://www.guideposts.org/blog/inexplicable-coincidence-on-a -hawaiian-beach.

3. For further details, see Pat Shellenbarger, "Man's Journey to Find Birth Mom Ends—at Work," *Seattle Times*, December 19, 2007, https://www.seattletimes.com /nation-world/mans-journey-to-find-birth-mom-ends-8212-at-work/. For a follow-up story, see also Shellenbarger, "Reunion of Man, Birth Mother Who Both Worked at Lowe's Enriches Lives of All," *Grand Rapids Press*, May 10, 2009, http:// www.mlive.com/news/grand-rapids/index.ssf/2009/05/reunion_of_man_birth_ mother_wh.html.

4. For details, as well as photographs of the objects, see Marija Georgievska, "Looking for a Hammer: The Largest Hoard of Roman Silver & Gold Found with a Metal Detector," *Vintage News*, November 19, 2016, https://www.thevintagenews.com/ 2016/11/19/looking-for-a-hammer-the-largest-hoard-of-roman-silver-gold-found -with-a-metal-detector/. See also Tommy Gee, "Eric Lawes Obituary," *Guardian* (London), July 23, 2015, https://www.theguardian.com/uk-news/2015/jul/23/ eric-lawes.

5. See the detailed timeline, prepared by the Atomic Heritage Foundation, at http:// www.atomicheritage.org/history/hiroshima-and-nagasaki-bombing-timeline.

6. See e.g. "Tsunami 2004," YouTube, uploaded July 1, 2017, https://www.youtube .com/watch?v=rRpAzsehLGA.

7. See e.g. Andrea Woo and Josh O'Kane, "Michael Buble Putting Career on Hold, Three-Year-Old Son Noah Has Cancer," *Globe and Mail* (Toronto), November 4, 2016, http://www.theglobeandmail.com/news/national/michael-buble-putting -career-on-hold-three-year-old-son-noah-has-cancer/article32674453/.

8. On December 20, 2017, *Us Weekly* reported that Buble's son is "doing well" with a "positive prognosis." See "Michael Buble Is 'Ready to Think About' Working Again After Son's Cancer Battle," *Us Weekly*, December 20, 2017, https://www.usmagazine .com/celebrity-news/news/michael-buble-ready-to-work-again-after-sons-cancer -battle/.

9. Leone is quoted on the movie review website Rotten Tomatoes (https://www .rottentomatoes.com/m/the_good_the_bad_and_the_ugly/).

10. See e.g. "Lottery Winner Dies Weeks after Cashing in $1 Million Scratch-Off Ticket," *Eyewitness News* (WLS-TV Chicago), January 30, 2018, http://abc7chicago .com/3008129.

11. A description of the "bitterly cold winter" of 1942–43 appears in Ron H. Pahl, *Breaking Away from the Textbook: Creative Ways to Teach World History*, vol. 2, *The Enlightenment through the 20th Century* (Lanham MD: ScarecrowEducation, 2002), 145. Pahl also writes that Hitler's "attack was carefully planned—his troops went to war with plenty of airplanes, tanks, and troops, but Hitler did not add one thing into his plans: the Russian winter." Similarly, the German army is described as "weakened by bitter cold and frost" prior to their January 1943 surrender in Jan Palmowski and Christopher Riches, *A Dictionary of Contemporary World History*, 4th ed. (Oxford: Oxford University Press, 2016).

12. A video of Roberts's comments, made at the Grade 9 graduation ceremony at Cardigan Mountain School in Canaan, New Hampshire, on June 3, 2017, accompanies an article on the *Washington Post* website. Robert Barnes, "The Best Thing Chief Justice Roberts Wrote This Term Wasn't a Supreme Court Opinion," *Washington Post*, July 2, 2017, https://www.washingtonpost.com/politics/courts _law/the-best-thing-chief-justice-roberts-wrote-this-term-wasnt-a-supreme -court-opinion/2017/07/02/b80a5afa-5e6e-11e7-9fc6-c7ef4bc58d13_story.html. For another report, which includes a full transcript of the speech, see Katie Reilly, "'I Wish You Bad Luck.' Read Supreme Court Justice John Roberts' Unconventional Speech to His Son's Graduating Class," *Time*, July 5, 2017, http://time.com/4845150/chief-justice-john-roberts-commencement-speech -transcript/.

13. See Joe Weisenthal, "We Love What Warren Buffett Says about Life, Luck, and Winning the 'Ovarian Lottery,'" *Business Insider*, December 10, 2013, http://www .businessinsider.com/warren-buffett-on-the-ovarian-lottery-2013-12/.

Chapter 4: The Day I Was Born

1. See e.g. Elisabeth Fraser, "Try Avoiding Bad Luck Today by Forgetting It's Friday the 13th," *CBC News*, January 13, 2017, http://www.cbc.ca/news/canada/new -brunswick/friday-the-13th-1.3932644.

2. See e.g. John Roach, "Friday the 13th Superstitions Rooted in Bible and More," *National Geographic*, May 14, 2011, https://news.nationalgeographic.com/news /2011/05/110513-friday-the-13th-superstitions-triskaidekaphobia/. Or see the colourful recounting "The 13th Guest," *Neatorama*, February 13, 2015, http://www .neatorama.com/2015/02/13/The-13th-Guest/.

3. See e.g. Dean Burnett, "Friday the 13th: Why Is It 'Unlucky'?," *Guardian* (London), February 13, 2015, https://www.theguardian.com/science/brain-flapping/2015/ feb/13/friday-13th-unlucky-why-science-psychology and the related discussion "Is Friday the 13th Really Unlucky?," *IFLScience*, http://www.iflscience.com/editors -blog/friday-13th-really-unlucky-day/.

4. Lawson's novel is available in a variety of digital formats, free of charge, at https:// www.gutenberg.org/files/12345/12345-h/12345-h.htm.

5. For historical discussions of the schooner *Thomas W. Lawson*, see e.g. Anthony F. Sarcone and Lawrence S. Rines, "A History of Shipbuilding at Fore River," *Quincy's Shipbuilding Heritage*, http://thomascranelibrary.org/shipbuildingheritage/history/ historyindex.html; "Thomas W. Lawson," http://www.fleetsheet.com/lawson.htm; and "My Favorite Schooner," Schooner Freedom Charters, June 14, 2014, http:// www.schoonerfreedom.com/my-favorite-schooner/.

6. See Clyde Haberman, "A Reading to Recall the Father of Tevye," *New York Times*, May 17, 2010, http://www.nytimes.com/2010/05/18/nyregion/18nyc.html.

7. See Fiammetta Rocco, "Ma'am Darling: The Princess Driven by Loyalty and Duty," *Independent* (London), February 25, 1998, https://www.independent. co.uk/news/maam-darling-the-princess-driven-by-loyalty-and-duty-1146783 .html.

8. The town is Port Dover, Ontario. See the event's website at http://www.pd13.com/ pages/1354815191/Origins. The "Friday the 13th Motorcycle Rally" page on Wikipedia (https://en.wikipedia.org/wiki/Friday_the_13th_motorcycle_rally) features several links to news reports.

9. See Laurie Ulster, "Happy Friday the 13th: Celebrities & Their Superstitions," Biography.com, April 11, 2018, http://www.biography.com/news/celebrity -superstitions.

10. See Joseph Ditta, "Friggatriskaidekaphobes Need Not Apply," *From the Stacks*, New York Historical Society, January 13, 2012, http://blog.nyhistory.org/friggatris -kaidekaphobes-need-not-apply/. See also Sadie Stein, "Morituri te Salutamus," *Paris Review*, March 13, 2015, https://www.theparisreview.org/blog/2015/03/13/ morituri-te-salutamus/.

11. See Hans Decos, "Friday the 13th: Numerology's Take on Luck and Superstition," Numerology.com, http://www.numerology.com/numerology-news/friday-the-13th -numerology.

12. See T.J. Scanlon, Robert N. Luben, F.L. Scanlon, and Nicola Singleton, "Is Friday the 13th Bad for Your Health?," *British Medical Journal* 307, no. 6919 (December 18, 1993): 1584–86, http://www.bmj.com/content/307/6919/1584. According to their Table I, the total traffic on the southern section of the M25 motorway was 1,283,853 over five Friday the 6ths, and 1,265,495 over five Friday the Thirteenths, for a ratio of 1,265,495 /1,283,853 = 0.9857 $\doteq$ 0.986 = 1 – 0.014, corresponding to a 1.4 percent decrease, which is statistically significant because of the large total numbers involved. And according to Table V, total hospital admissions in the South West Thames region due to transport accidents was 45 over six Friday the 6ths, and 65 over six Friday the Thirteenths, for a ratio of 65/45 = 1.4444 $\doteq$ 1.44, corresponding to a 44 percent increase. They claim this last result is statistically significant ("$p < 0.05$"), but using "poisson.test(c(45,65))" in the free statistical software package R (available at https://cran.r-project.org/), I actually compute a p-value of 0.06957, which is greater than 0.05, indicating it is actually not quite statistically significant.

13. See Remy Melina, "Statistically Speaking, Is Friday the 13th Really Unlucky?," *LiveScience*, January 13, 2012, https://www.livescience.com/17900-statistically -speaking-friday-13th-unlucky.html.

14. The Christmas issue of the *British Medical Journal* is discussed on the journal's website here: http://www.bmj.com/about-bmj/resources-authors/article-types/ christmas-issue. For a retrospective survey of *BMJ* Christmas papers, see Navjoyt Ladher, "Christmas Crackers: Highlights from Past Years of *The BMJ*'s Seasonal Issue," *British Medical Journal* 355, no.8086 (December 17, 2016): i6679, http:// www.bmj.com/content/355/bmj.i6679.

15. Simo Nähyä. "Traffic Deaths and Superstition on Friday the 13th," *American Journal of Psychiatry* 159, no. 12 (December 2002): 2110–11, http://ajp .psychiatryonline.org/doi/abs/10.1176/appi.ajp.159.12.2110.

16. Igor Radun and Heikki Summala, "Females Do Not Have More Injury Road Accidents on Friday the 13th," *BMC Public Health* 4 (2004), https://bmcpublichealth .biomedcentral.com/articles/10.1186/1471-2458-4-54.

17. See "Friday 13th Not More Unlucky, Study Shows," Reuters, June 12, 2008, http:// www.reuters.com/article/us-luck-odd-idUSHER25778420080612.

18. See the comment by "Datacharmer" quoted at http://andrewgelman.com/2008/ 08/22/friday_the_13th_1/.

19. Aristomenis K. Exadaktylos et al., "Friday the 13th and Full-Moon: The 'Worst Case Scenario' or only Superstition?," *American Journal of Emergency Medicine* 19, no. 4 (July 2001): 319–20.

20. Melanie Wright, "Coincidence? 13 Really Is the Unlucky Number," *Telegraph* (London), November 19, 2005, http://www.telegraph.co.uk/finance/personal finance/2926352/Coincidence-13-really-is-the-unlucky-number.html.

21. See the National Lottery Lotto Number Frequency Table available at https://www .lottery.co.uk/lotto/statistics.

22. Based on the draw period "Oct 2015 to Present" as retrieved on August 2, 2017, from https://www.lottery.co.uk/lotto/statistics. The overall average is found by

taking the total number of draws included (189), multiplying by the number of balls drawn each time (six), and dividing by the number of available numbers (59) to obtain $189 \times 6 / 59 = 19.22034 \doteq 19.2$.

23. See e.g. the Powerball Frequency Chart at https://www.ctlottery.org/Modules/FCharts/default.aspx?id=5.

24. Based on data from the Lotto 6/49 Statistics website ("Overall 1982 to Present without Bonus") available at http://www.lotto649stats.com/overall_frequency.html, retrieved on August 2, 2017.

25. For Powerball, using "poisson.test(11,13.84)" in R gives a p-value of 0.5887. For Lotto 6/49, using "poisson.test(402,428.3)" gives a p-value of 0.209. So, both p-values are well above 0.05, indicating no statistical significance.

Chapter 5: Our Love of Magic

1. See Richard Webster, *The Encyclopedia of Superstitions* (Woodbury, MN: Llewellyn Publications, 2008).

2. See "Special Interviews with M. Night Shamylan [*sic*] and the Cast of *The Sixth Sense*," YouTube, uploaded February 16, 2014, https://www.youtube.com/watch?v=cFuqtlTmNlA&t=15m19s.

3. According to an estimate of the total worldwide gross from November 4, 2016, to March 16, 2017, on the Box Office Mojo website. "*Doctor Strange* (2016)," Box Office Mojo, http://www.boxofficemojo.com/movies/?id=marvel716.htm.

4. His name is Owen Anderson ("fun-filled magic shows for kids and families across southern Ontario"), and his website is http://www.owenanderson.ca.

5. Alexis Cheung, "In Pursuit of Ghosts," *re:Porter*, October 2016, 22–31, https://static.flyporter.com/Content/reporter/54.pdf.

6. Brett J. Talley was nominated in November 2017 by President Donald Trump to the Federal District Court in Montgomery, Alabama. Talley was apparently part of the Tuscaloosa Paranormal Research Group between 2009 and 2010 and wrote books about paranormal activities. See e.g. Gideon Resnick and Sam Stein, "Before He Was Tapped by Donald Trump, Controversial Judicial Nominee Brett J. Talley Investigated Paranormal Activity," *Daily Beast*, November 13, 2017, https://www.thedailybeast.com/before-he-was-nominated-for-federal-court-donald-trumps-controversial-judicial-nominee-brett-j-talley-hunted-ghosts.

7. See e.g. Sammy Said, "The Top 10 Most Sold Board Games Ever," TheRichest.com, https://www.therichest.com/rich-list/most-popular/the-top-10-most-sold-board-games-ever/.

8. Aaron Boone was traded from the Cincinnati Reds to the Boston Red Sox on July 31, 2003, and he hit his 11th-inning home run against the Yankees on October 16, 2003.

9. See e.g. Jill Martin, "Believe It! Chicago Cubs End the Curse, Win 2016 World Series," CNN, November 3, 2016, http://www.cnn.com/2016/11/02/sport/world-series-game-7-chicago-cubs-cleveland-indians/.

10. Post on Humans of New York website, http://www.humansofnewyork.com/post/151386313471/id-been-harboring-a-crush-on-him-since-5th.

11. "Serendipity Elevator," YouTube, uploaded June 15, 2007, https://www.youtube.com/watch?v=rrvKt7GNSco.

12. Details on the film, *My Year without Sex*, written and directed by Sarah Watt, can be found at http://www.imdb.com/title/tt1245358/. The lottery ticket is mentioned briefly in Linda Burgess, "Film Review: My Year Without Sex," Stuff.com, November 13, 2009, http://www.stuff.co.nz/entertainment/film/film-reviews/3015317/Film-review-My-Year-Without-Sex.

13. See e.g. Brandi Reissenweber, "What's This Business about 'Chekhov's Gun'?," *Gotham Writers*, https://www.writingclasses.com/toolbox/ask-writer/whats-this-business-about-chekhovs-gun.

Chapter 7: Luck Revisited

1. See Ashifa Kassam, "Canadian Woman, 84, Finds Long-Lost Diamond Ring Wrapped around Carrot," *Guardian* (London), August 16, 2017, https://www.theguardian.com/world/2017/aug/16/canadian-woman-engagement-ring-carrot.

2. The probability of choosing four numbers correctly out of six numbers from 1 to 49 is $\binom{6}{4} \times \binom{43}{2} / \binom{49}{6} \doteq 0.0009686197$, or just under one chance in 1,000.

3. See e.g. "Total Number of Existing and New Car Models Offered in the U.S. Market from 2000 to 2017," Statista.com, https://www.statista.com/statistics/200092/total-number-of-car-models-on-the-us-market-since-1990/.

Chapter 8: Lucky News

1. Fiona Mathews, Paul J. Johnson, and Andrew Neil, "You Are What Your Mother Eats: Evidence for Maternal Preconception Diet Influencing Foetal Sex in Humans," *Proceedings of the Royal Society B: Biological Sciences* 275, no. 1643 (July 22, 2008): 1661–68, http://rspb.royalsocietypublishing.org/content/275/1643/1661.

2. See e.g. "You Are What Your Mother Eats," University of Exeter, April 25, 2008, http://www.exeter.ac.uk/news/archive/2008/april/title_626_en.html.

3. See e.g. Amy A. Gelfand, "Acupuncture for Migraine Prevention: Still Reaching for Convincing Evidence," *Journal of the American Medical Association: Internal Medicine* 177, no. 4 (April 2017): 516–17, https://jamanetwork.com/journals/jamainternalmedicine/article-abstract/2603487.

4. See Ricardo de la Vega et al., "Induced Beliefs about a Fictive Energy Drink Influences 200-m Sprint Performance," *European Journal of Sport Science* 17, no. 8 (2017): 1084–89, https://www.ncbi.nlm.nih.gov/pubmed/28651483.

5. See e.g. S. Stanley Young, Heejung Bang, and Kutluk Oktay, "Cereal-Induced Gender Selection? Most Likely a Multiple Testing False Positive," *Proceedings of the Royal Society B: Biological Sciences* 276, no. 1660 (April 7, 2009): 1211–12), http://rspb.royalsocietypublishing.org/content/276/1660/1211. See also "Study Refutes Notion that Eating a Certain Cereal Will Result in More Male Babies," *ScienceDaily*, January 17, 2009, https://www.sciencedaily.com/releases/2009/01/090114075759.htm.

6. See e.g. "Teen Saves Life of Woman Who Saved Him," NBC News, February 5, 2006, http://www.nbcnews.com/id/11190559/ns/us_news-weird_news/t/teen-saves-life-woman-who-saved-him/.

7. See e.g. the photograph of an old Connecticut billboard with this slogan in the Duke University Libraries Digital Collections, http://library.duke.edu/digitalcollections/oaaaarchives_BBB2576/.

8. The story is available at http://faculty.smu.edu/nschwart/2312/lifeyousave.htm.

9. An episode of the Schlitz Playhouse of Stars anthology series, entitled simply "The Life You Save" (http://www.imdb.com/title/tt0394872/).

10. See the excerpt at https://www.youtube.com/watch?v=yQVYQET0Yh0.

11. According to the American Heart Association (http://www.heart.org/HEARTORG/CPRAndECC/WhatisCPR/CPRFactsandStats/CPR-Statistics_UCM_307542_Article.jsp), each year in the United States there are about 350,000 out-of-hospital cardiac arrests, of which about 46 percent get immediate help, of which about 25 percent survive. So, the number of "cardiac arrest saved by CPR" incidents each year (not even counting other ways CPR can save lives, such as the Heimlich manoeuvre) is about $350,000 \times 0.46 \times 0.25 \doteq 40,000$. Also, nearly 100 million Americans (nearly 30 percent) know CPR, so the chance that one specific American who knows CPR will save a life this year is about 40,000 / 100,000,000, or one in 2,500. Let us imagine that Americans are divided into 6,000 local "communities" (either small towns, or distinct social groups within larger cities) of about 50,000 people each. Then, if A and B are in the same community and both know CPR, the chance that A will save B's life this year is about one chance in $2,500 \times 50,000 = 1.25 \times 10^8$. So the chance they will save each other's life this year is about one in $(1.25 \times 10^8)^2 \doteq 1.5 \times 10^{16}$. If we choose two Americans who know CPR completely at random, they have one chance in 6,000 of being in the same community, so their overall chance of saving each other's lives this year is about one in $1.5 \times 10^{16} \times 6,000 = 9 \times 10^{19}$, or one chance in 90 billion billion.

12. Continuing the previous calculation, and assuming I will live a total of 80 years, if A and B each live in the same community and each know CPR, then the probability that A saves B's life at some point during my lifetime is about 80 / $(1.25 \times 10^8) \doteq 1$ / (1.5×10^6), or about one chance in 1.5 million. So, the chance they will *each* save each other's life during my lifetime is about one chance in $(1.5 \times 10^6)^2 = 2.25 \times 10^{12}$. Furthermore, each community has about $50,000 \times 30\% = 15,000$ people who know CPR, so the number of such pairs of A and B in each community is about $\binom{15,000}{2} \doteq (15,000)^2 / 2 = 112,500,000$. Since there are about 6,000 communities, we conclude that the probability of the existence of *some* such pair who save each other's lives during my lifetime is about $6,000 \times 112,500,000 / (2.25 \times 10^{12})$ which equals 0.3, or about 1/3.

13. See "Most Lightning Strikes Survived," Guinness World Records, http://www.guinnessworldrecords.com/world-records/most-lightning-strikes-survived.

14. See also the explicit descriptions of his early encounters in Hank Burchard, "Lightning Strikes 4 Times," *Lakeland Ledger*, May 2, 1972, https://news.google

.com/newspapers?nid=1347&dat=19720502&id=OScVAAAAIBAJ&sjid
=afoDAAAAIBAJ&pg=7465,354926.

15. According to the National Lightning Safety Institute, based on data compiled
by Ronald L. Holle. See "Lightning Deaths in the United States—Weighted by
Population, 1990 to 2003," National Lightning Safety Institute, http://lightningsafety
.com/nlsi_lls/fatalities_us.html.

16. See e.g. "Celebrities Talk, 'The Law of Attraction' (So Inspiring!)," YouTube,
uploaded December 19, 2014, https://www.youtube.com/watch?v=xfSLm7swfp4;
"Oprah Winfrey Speaks about the Secret—Law of Attraction and How to Use It!,"
YouTube, uploaded May 9, 2013, https://www.youtube.com/watch?v=-Zm3
-exDWIg; and "Steve Harvey Talks about the Law of Attraction . . . It Works!,"
YouTube, uploaded March 27, 2013, https://www.youtube.com/watch?v
=zrE7dq1b9fc.

17. See the transcript and video on the *New York Times* website. Giovanni Russonello,
"Read Oprah's Golden Globes Speech," *New York Times*, January 7, 2018, https://
www.nytimes.com/2018/01/07/movies/oprah-winfrey-golden-globes-speech
-transcript.html. A video can also be found at Golden Globes, "Oprah Winfrey
Receives the Cecil B. deMille Award—Golden Globes 2018," YouTube, uploaded
January 7, 2018, https://www.youtube.com/watch?v=LyBims8OkSY.

18. See "Cancer Angel on the Couch 03/07," YouTube, uploaded September 1, 2009,
https://www.youtube.com/watch?v=7uf-5yuRiPs. See also Weston Kosova, "Why
Health Advice on 'Oprah' Could Make You Sick," *Newsweek*, May 29, 2009, http://
www.newsweek.com/why-health-advice-oprah-could-make-you-sick-80201; Bart B.
van Bockstaele, "Kim Tinkham, the Woman Whom Oprah Made Famous, Dead at
53," *Digital Journal*, December 8, 2010, http://www.digitaljournal.com/article/
301197; Beatis, "Alternative Medicine: Double Corruption," *Anaximperator Blog*,
February 23, 2011, https://anaximperator.wordpress.com/2011/02/23/alternative-
medicine-double-corruption/; and Beatis, "Quack Victim Kim Tinkham Dies of
Breast Cancer," *Anaximperator Blog*, December 8, 2010, https://anaximperator
.wordpress.com/2010/12/08/orac-of-respectful-insolence-just-announced-that-kim
-tinkham-has-died-of-breast-cancer/.

19. Yilun Wang and Michal Kosinski, "Deep Neural Networks Are More Accurate than
Humans at Detecting Sexual Orientation from Facial Images," *OSFHome*, February
15, 2017, https://osf.io/zn79k/.

20. See "Advances in AI Are Used to Spot Signs of Sexuality," *Economist*, September 9,
2017, https://www.economist.com/news/science-and-technology/21728614
-machines-read-faces-are-coming-advances-ai-are-used-spot-signs and "Row over
AI that 'Identifies Gay Faces,'" BBC News, September 11, 2017, http://www.bbc
.com/news/technology-41188560. See also the follow-up study: Blaise Agüera y
Arcas, Alexander Todorov, and Margaret Mitchell, "Do Algorithms Reveal Sexual
Orientation or Just Expose Our Stereotypes?," *Medium*, https://medium.com
/@blaisea/do-algorithms-reveal-sexual-orientation-or-just-expose-our-stereotypes
-d998fafdf477.

21. Professor Benedict Jones of the University of Glasgow, quoted in the BBC article cited above.

22. Alec T. Beall and Jessica L. Tracy, "Women Are More Likely to Wear Red or Pink at Peak Fertility," *Psychological Science* 24, no. 9 (September 2013): 1837–41, http://ubc-emotionlab.ca/wp-content/files_mf/bealandtracypsonlinefirst.pdf.

23. See e.g. Rachael Rettner, "Fertile Women More Likely to Wear Red," *LiveScience*, May 28, 2013, https://www.livescience.com/34737-fertile-peak-women-wear-red.html; Seriously Science, "Women Are More Likely to Wear Red or Pink at Peak Fertility," *Discover*, July 22, 2013, http://blogs.discovermagazine.com/seriously-science/2013/07/22/women-are-more-likely-to-wear-red-or-pink-at-peak-fertility/#.WaltiN8QTmE; and Justin Lehmiller, "Women Reach for Red and Pink Clothes during Ovulation," Sex & Psychology by Dr. Justin Lehmiller, July 31, 2013, http://www.lehmiller.com/blog/2013/7/31/women-reach-for-red-and-pink-clothes-during-ovulation.

24. Andrew Gelman, "Too Good to Be True," *Slate*, July 24, 2013, http://www.slate.com/articles/health_and_science/science/2013/07/statistics_and_psychology_multiple_comparisons_give_spurious_results.html.

25. Jessica Tracy and Alec Beall, "Too Good Does Not Always Mean Not True," UBC Emotion and Self Lab, July 30, 2013, http://ubc-emotionlab.ca/2013/07/too-good-does-not-always-mean-not-true/.

26. Catriona Harvey-Jenner, "So THIS Is Why Women's Periods Tend to Sync Up," *Cosmopolitan*, July 25, 2016, http://www.cosmopolitan.com/uk/body/health/news/a44886/why-womens-periods-sync-up/.

27. Martha K. McClintock, "Menstrual Synchrony and Suppression," *Nature* 229 (1971): 244–45.

28. See e.g. H. Clyde Wilson, "A Critical Review of Menstrual Synchrony Research," *Psychoneuroendocrinology* 17, no. 6 (November 1992): 565–91.

29. See e.g. H. Clyde Wilson, Sarah Hildebrandt Kiefhaber, and Virginia Gravel, "Two Studies of Menstrual Synchrony: Negative Results," *Psychoneuroendocrinology* 16, no. 4 (1991): 353–59; Beverly I. Strassmann, "Menstrual Synchrony Pheromones: Cause for Doubt," *Human Reproduction* 14, no. 3 (March 1999): 579–80, https://academic.oup.com/humrep/article/14/3/579/632869/Menstrual-synchrony-pheromones-cause-for-doubt; and Zhengwei Yang and Jeffrey C. Schank, "Women Do Not Synchronize Their Menstrual Cycles," *Human Nature* 17, no. 4 (December 2006): 433–47.

30. Beverly I. Strassmann, "The Biology of Menstruation in Homo Sapiens: Total Lifetime Menses, Fecundity, and Nonsynchrony in a Natural-Fertility Population," *Current Anthropology* 38, no. 1 (February 1997): 123–29. Strassmann concluded that "the null hypothesis that the women's menstrual onsets were independent cannot be rejected."

31. See e.g. Anna Gosline, "Do Women Who Live Together Menstruate Together?," *Scientific American*, December 7, 2007, https://www.scientificamerican.com/article/do-women-who-live-together-menstruate-together/; Luisa Dillner, "Do Women's

Periods Really Synchronise When They Live Together?," *Guardian* (London), August 15, 2016, https://www.theguardian.com/lifeandstyle/2016/aug/15/periods -housemates-menstruation-synchronise; and Charlotte McDonald, "Is It True that Periods Synchronise When Women Live Together?," BBC News, September 7, 2016, http://www.bbc.com/news/magazine-37256161.

32. See M.A. Arden, L. Dye, and A. Walker, "Menstrual Synchrony: Awareness and Subjective Experiences," *Journal of Reproductive and Infant Psychology* 17, no. 3 (1999): 255–65, http://www.tandfonline.com/doi/abs/10.1080/02646839908404593 and Breanne Fahs, "Demystifying Menstrual Synchrony: Women's Subjective Beliefs about Bleeding in Tandem with Other Women," *Women's Reproductive Health* 3, no. 1 (2016): 1–15.

33. For example, the Cleveland Clinic (https://my.clevelandclinic.org/health/articles /10132-normal-menstruation) writes, "Most women bleed for 3 to 5 days, but a period lasting only 2 days to as many as 7 days is still considered normal."

34. Alexandra Alvergne, associate professor in biocultural anthropology at the University of Oxford, as quoted in McDonald, "Is It True that Periods Synchronise When Women Live Together?"

Chapter 9: Supremely Similar

1. Sima Sahar Zerehi, "'I Won the DNA Lottery': Woman Finds Biological Father after Lifelong Search," CBC News, August 7, 2016, http://www.cbc.ca/news/canada /north/nunavut-woman-finds-biological-father-1.3709705.

2. Marsha Lederman, "Two Friends in Calgary Discover They Are Really Long-Lost Brothers," *Globe and Mail* (Toronto), July 26, 2013, https://beta.theglobeandmail .com/news/national/at-50-and-46-friends-discover-they-are-really-long-lost -brothers/article13469118/.

3. According to ChildTrends.com (https://www.childtrends.org/indicators/adopted -children), about 2 percent of US children are adopted. According to Lee Helland, "Level of Involvement for Birth Parents," Parents.com, https://www.parents.com/ parenting/adoption/parenting/level-of-involvement-for-birth-parents, 36 percent of adoptees have some contact with their birth families, corresponding to approximately 323,000,000 × 0.02 × 0.36, or about 2.3 million adoptees. A paper published in 2002 notes that in a single month in the year 2000 in the state of Oregon alone, 2,529 adoptees requested birth records, of whom about 15 percent (33 of 221 in their sample), or 379 of them, successfully found their birth mother. Julia C. Rhodes et al., "Releasing Pre-Adoption Birth Records: A Survey of Oregon Adoptees," *Public Health Reports* 117, no. 5 (September/October 2002): 463–71.

4. See many discussions, including: "Lincoln–Kennedy Coincidences Urban Legend," Wikipedia, https://en.wikipedia.org/wiki/Lincoln-Kennedy_coincidences_urban_ legend; David Mikkelson, "Lincoln and Kennedy Coincidences," Snopes, October 30, 2017, https://www.snopes.com/fact-check/linkin-kennedy/; Bruce Martin, "Coincidences: Remarkable or Random?," *Skeptical Inquirer* 22, no. 5 (September/ October 1998), http://www.csicop.org/si/show/coincidences_remarkable_or_random;

Ron Kurtus, "Similarities between the Assassinations of Kennedy and Lincoln (1860s and 1960s)," Ron Kurtus' School for Champions, July 10, 2015, http://www.school-for-champions.com/history/lincolnjfk.htm; and Brian Galindo, "10 Weird Coincidences between Abraham Lincoln and John F. Kennedy," *Buzzfeed*, https://www.buzzfeed.com/briangalindo/10-weird-coincidences-between-abraham-lincoln-and-john-f-ken.

5. See e.g. the list of 1,184 such surnames at Behind the Name (https://surnames.behindthename.com/names/length/7).

Chapter 11: Protected by Luck

1. For example, suppose 2 percent of people are thieves, and 2 percent are so helpful and outgoing that they will spontaneously and honestly offer to help you. If you choose a random person, there is only a 2 percent chance they are a thief. But among those who offer to help, fully half (50 percent) are thieves—a much higher fraction.

2. See Carrie Miller, "Six Things Solo Travel Teaches You," *National Geographic*, August 24, 2016, http://www.nationalgeographic.com/travel/travel-interests/tips-and-advice/six-things-solo-alone-travel-teaches-you/.

3. As quoted in Tom Jackman, "Trump Makes False Statement about US Murder Rate to Sheriffs' Group," *Washington Post*, February 7, 2017, https://www.washingtonpost.com/news/true-crime/wp/2017/02/07/trump-makes-false-statement-about-u-s-murder-rate-to-sheriffs-group/.

4. The complete text is available at "The Inaugural Address," White House, January 20, 2017, https://www.whitehouse.gov/inaugural-address.

5. See e.g. "United States Crime Rates 1960–2016" Disaster Center, http://www.disastercenter.com/crime/uscrime.htm; see also "Crime in the United States—By Volume and Rate per 100,000 Inhabitants, 1993–2012," FBI Uniform Crime Reporting Program, https://ucr.fbi.gov/crime-in-the-u.s/2012/crime-in-the-u.s.-2012/tables/1tabledatadecoverviewpdf/table_1_crime_in_the_united_states_by_volume_and_rate_per_100000_inhabitants_1993-2012.xls and "Crime in the United States—By Volume and Rate per 100,000 Inhabitants, 1997–2016," FBI Uniform Crime Reporting Program, https://ucr.fbi.gov/crime-in-the-u.s/2016/crime-in-the-u.s.-2016/tables/table-1.

6. See e.g. "A Look Back to 2005: 'The Summer of the Gun,'" Global News, June 8, 2016, http://globalnews.ca/video/2750283/a-look-back-to-2005-the-summer-of-the-gun.

7. Detective Sergeant Savas Kyriacou, as quoted in "'Toronto Has Lost Its Innocence,' Police Say of Boxing Day Shooting," CBC News, December 27, 2005, http://www.cbc.ca/news/canada/toronto-has-lost-its-innocence-police-say-of-boxing-day-shooting-1.569480.

8. Christina Blizzard, "Turning Murder into Politics," *Toronto Sun*, May 25, 2007.

9. In the city of Toronto, there were 64 homicides in 2004 and 80 in 2005, and 80 / 64 = 1.25. With a city population of about 2.5 million, this corresponds to homicide rates per 100,000 residents of about 2.5566 in 2004 and 3.1958 in 2005. See e.g. page 2 of

"Gangs and Guns," Coalition for Gun Control, https://www.webcitation.org/
5tYeuI09Y?url=http://www.guncontrol.ca/English/Home/Works/gangsandguns
.pdf and "Homicide," Toronto Police Service Public Safety Data Portal, http://data
.torontopolice.on.ca/datasets/homicide?orderBy=Occurrence_year.

10. There were 89 homicides in the city of Toronto in 1991, and a population of about
2.28 million, hence a rate of about 3.9108; see e.g. the above Coalition for Gun
Control report, or Alek Gazdic, "Despite Rise, Police Say T.O. Murder Rate 'Low,'"
CTV News, December 26, 2007, http://www.ctvnews.ca/despite-rise-police-say
-t-o-murder-rate-low-1.268936 and Betsy Powell, "Toronto Police 'Struggling'
to Solve Murders," *Toronto Star*, January 1, 2011, https://www.thestar.com/news/
crime/2011/01/01/toronto_police_struggling_to_solve_murders.html.

11. For example, FBI statistics show that in 2005, New York City had 539 murders
and non-negligent manslaughters in a population of 8,115,690, corresponding to a
rate of 539 / 8,115,690 3 100,000 = 6.64; the city of Los Angeles had 489 in a population of 3,871,077 for a rate of 489 / 3,871,077 3 100,000 = 12.63; the city of Atlanta
had 90 in a population of 430,666 for a rate of 90 / 430,666 3 100,000 = 20.90; the
city of Detroit had 354 in a population of 900,932 for a rate of
354 / 900,932 3 100,000 = 39.29. "Table 6: Crime in the United States by
Metropolitan Statistical Area, 2005," FBI Uniform Crime Reporting Program,
https://www2.fbi.gov/ucr/05cius/data/table_06.html.

12. For example, according to Statistics Canada, in 2005 the Winnipeg census
metropolitan area had a homicide rate of 3.72, the Regina census metropolitan
area had a homicide rate of 3.96, and the Edmonton census metropolitan area
had a homicide rate of 4.19. See "Table 253-0004: Homicide Survey, Number and
Rates (per 100,000 Population) of Homicide Victims, by Census Metropolitan
Area (CMA)," Statistics Canada, http://www5.statcan.gc.ca/cansim/a26?lang
=eng&retrLang=eng&id=2530004&&pattern=&stByVal=1&p1=1&p2=-1&
tabMode=dataTable&csid=.

13. According to Statistics Canada's CANSIM Table 253-0004, in 2005 there were 2.06
homicides per 100,000 population in Canada as a whole, and just 1.98 per 100,000
in the Toronto census metropolitan area.

14. See e.g. "A Sequel to 'Summer of the Shark'?," CNN, May 22, 2002, http://www
.cnn.com/2002/US/05/21/shark.attacks/ and Cathy Keen, "'Summer of the Shark'
in 2001 More Hype Than Fact, New Numbers Show," University of Florida, http://
news.ufl.edu/archive/2002/02/summer-of-the-shark-in-2001-more-hype-than-fact
-new-numbers-show.html. Worldwide, there were 76 unprovoked shark attacks in
2001, leading to five deaths—fewer than the 85 attacks and 12 deaths the previous
year. However, there were three such deaths in the United States in 2001, compared
to just one in 2000 and none in 2002.

15. According to the National Safety Council's estimate at https://www.nsc.org/road
-safety/safety-topics/fatality-estimates, there were 40,327 motor vehicle deaths in
2016 in the United States, compared to 17,250 murders according to the Disaster

Center study cited in note 5. Therefore, there were 2.34 times as many motor vehicle deaths as murders.

16. From the data in the Toronto Police Service's Public Safety Data Portal, the numbers of homicides in the city of Toronto by year from 2005 through 2016 were: 80, 70, 86, 70, 62, 63, 51, 56, 57, 58, 57, and 74. Hence, in every year from 2006 through 2016 there were fewer than the 80 in 2005—significantly fewer in some years (for example, 2011 had 51, a decrease of just over 36 percent, since 51/80 is just under 0.64). And for 2006, 70/80 = 0.875, corresponding to a 12.5 percent decrease.

17. For example, according to page 12 of the 2011 Annual Statistical Report of the Toronto Police Service, available at http://www.torontopolice.on.ca/publications/files/reports/2011statsreport.pdf, a total of 3,139 bicycles were stolen in the city of Toronto in the year 2011.

18. See e.g. the data and graphs at http://www.planecrashinfo.com/cause.htm.

19. For example, the US Bureau of Transportation Statistics indicates there were 928,900,000 airline passengers in the United States in the year 2016. "Annual Passengers on All U.S. Scheduled Airline Flights (Domestic & International) and Foreign Airline Flights to and from the United States, 2003-2016," Bureau of Transportation Statistics, https://www.bts.gov/content/annual-passengers-all-us-scheduled-airline-flights-domestic-international-and-foreign.

20. See e.g. "TO70's Civil Aviation Safety Review 2016," available at http://to70.com/safety-review-2016/.

21. One source gives the odds of being killed on a single airline flight as 1 in 29.4 million. "Airplane Crash Statistics," Statistic Brain, http://www.statisticbrain.com/airplane-crash-statistics/.

22. See e.g. "Major Airlines Arrival Performance," Tableau Public, https://public.tableau.com/profile/flightstats#!/vizhome/AirlineMonthlyOTP2014/MajorAirlinesbyRegion.

23. As my friend realized, approximately two deaths per year in the United States are the result of bear attacks, compared to five from snakes, 48 from bees and wasps, and over 3,500 from drowning. See e.g. Mike Rogers, "Bear Attacks—Killer Statistic That May Surprise You," *Alaska Life*, July 2017, https://www.thealaskalife.com/outdoors/bear-attacks-statistic/.

24. According to estimates based on data from the Federation of American Scientists, the worldwide inventory reached a high of 64,449 nuclear weapons in 1986, before being reduced to 10,145 in 2014 (the latest year available). Max Roser and Mohamed Nagdy, "Nuclear Weapons," Our World in Data, https://ourworldindata.org/nuclear-weapons.

25. The Hiroshima and Nagasaki bombs were equivalent to about 15 and 21 kilotons (not megatons) of TNT, respectively. So, a four-megaton bomb is about 267 times as powerful as the Hiroshima bomb, and 190 times as powerful as the one dropped on Nagasaki.

26. A photo can be found at https://upload.wikimedia.org/wikipedia/commons/b/b0/ Goldsboro_Mk_39_Bomb_1-close-up.jpeg.

27. See Gary Hanauer, "The Story behind the Pentagon's Broken Arrows," *Mother Jones*, April 1981, https://books.google.ca/books?id=tOYDAAAAMBAJ. The discussion of the number of interlocks can be found on page 28. See also "Brush with Catastrophe?," *Full Story*, December 10, 2000, http://www.ibiblio.org/bomb/brush.html.

28. "Information Supplied by Nuclear Weapons Historian Chuck Hansen," *Full Story*, http://www.ibiblio.org/bomb/hansen_doc.html.

29. The Pentagon's narrative states that the bomb "fell free and broke apart upon impact." "First Things First: It Did Happen," *Full Story*, http://www.ibiblio.org/ bomb/initial.html.

30. See Keith Sharon, "Orange Resident Recalls Holding Future in His Hands," *Orange County Register*, December 31, 2012, http://www.ocregister.com/2012/12/31/ orange-resident-recalls-holding-future-in-his-hands/. The first six steps were described as: 1. the arming wires had been pulled; 2. the pulse generators had been activated; 3. the explosive actuators had been fired; 4. timers had started; 5. barometric switches had been engaged; and 6. low-voltage batteries were actuated. The seventh was a separate ARM/SAFE switch. Strangely, when this switch was later found, it was apparently set to ARM. This led to confusion—apparently still unresolved—about why the bomb didn't explode after all.

31. The two-page memo was written on October 22, 1969, by Parker F. Jones, supervisor of the nuclear weapons safety department at Sandia National Laboratories, in response to (and in disagreement with) the book *Kill and Overkill* by Ralph Lapp. It was titled "Goldsboro Revisited, or, How I Learned to Mistrust the H-Bomb, or, To Set the Record Straight." A digitized version of the document can be found on the *Guardian*'s website at https://www.theguardian.com/world/interactive/2013/ sep/20/goldsboro-revisited-declassified-document.

32. A writer for the *Guardian* wrote in 2013, "Had the device detonated, lethal fallout could have been deposited over Washington, Baltimore, Philadelphia and as far north as New York city—putting millions of lives at risk." Ed Pilkington, "US Nearly Detonated Atomic Bomb over North Carolina—Secret Document," *Guardian* (London), September 20, 2013, https://www.theguardian.com/ world/2013/sep/20/usaf-atomic-bomb-north-carolina-1961.

33. Joseph Wilson, "Random Acts," *NOW* (Toronto), November 17, 2005, https:// nowtoronto.com/art-and-books/books/random-acts/.

34. FBI crime statistics show clear downward trends in the rates of murder and violent crime. See e.g. "Crime in the United States—By Volume and Rate per 100,000 Inhabitants, 1997–2016," FBI Uniform Crime Reporting Program (see note 5).

35. The talk was hosted by the Perimeter Institute for Theoretical Physics in Waterloo, Ontario, on April 2, 2008. See "Jeffrey Rosenthal Probability Lecture at the Perimeter Institute," YouTube, uploaded November 25, 2010, https://www.youtube .com/watch?v=hWp6SBr_ZYU.

Chapter 12: Statistical Luck

1. "As the world becomes more quantitative and data-focused, mathematics takes center stage, with Statistician topping the best jobs of 2017." See "The Best Jobs of 2017," CareerCast, http://www.careercast.com/jobs-rated/best-jobs-2017.

2. See "U.S. News & World Report Announces the 2017 Best Jobs," *U.S. News & World Report*, https://www.usnews.com/info/blogs/press-room/articles/2017-01-11/us-news-announces-the-2017-best-jobs. The magazine also ranked statistician as the fourth-best job *overall* (after dentist, nurse practitioner, and physician assistant). See "The 100 Best Jobs," *U.S. News & World Report*, https://money.usnews.com/careers/best-jobs/rankings/the-100-best-jobs.

3. See e.g. Julie Rehmeyer, "Florence Nightingale: The Passionate Statistician," *Science News*, November 26, 2008, https://www.sciencenews.org/article/florence-nightingale-passionate-statistician and Eileen Magnello, "Florence Nightingale: The Compassionate Statistician," *Plus Magazine*, December 8, 2010, https://plus.maths.org/content/florence-nightingale-compassionate-statistician. A higher-resolution image of Nightingale's famous polar area diagram of causes of mortality in the army is available at https://upload.wikimedia.org/wikipedia/commons/1/17/Nightingale-mortality.jpg. And a photograph of the form nominating Nightingale for membership in the Royal Statistical Society is posted at https://twitter.com/HetanShah/status/940195192342237189.

4. The survey data is still available on my website (http://probability.ca/jeff/teaching/1617/sta130/studentdata.txt). There were 80 students in total, consisting of 41 males (of whom 14, or 34 percent, were in a romantic relationship, while 27 were not) and 39 females (of whom 11, or 28 percent, were, while 28 were not).

5. I used the standard normal-approximation t-test for comparisons of proportions, via the R command "prop.test(matrix(c(14,11,27,28), nrow=2))." This test returned a p-value of 0.7401, and a 95 percent confidence interval of (-0.168, 0.287), meaning that the males could plausibly have between 28.7 percent more relationships and 16.8 percent fewer.

6. I used the standard normal-approximation t-test for comparisons of means, via the R command "t.test()." This test returned a p-value of $8.479e-11$, which equals $1 / 11{,}793{,}677{,}973$, and gave a 95 percent confidence interval of (9.18, 15.68) centimetres, or about (3.6, 6.2) inches, meaning that, on average, males could plausibly be between 3.6 and 6.2 inches taller than females.

7. I thank Louella Lobo for encouraging me to discuss this aspect, and for all of her support.

8. Michelle M. Stein et al., "Innate Immunity and Asthma Risk in Amish and Hutterite Farm Children," *New England Journal of Medicine* 375, no. 5 (August 4, 2016): 411–21, http://www.nejm.org/doi/full/10.1056/NEJMoa1508749.

9. Eric L. Simpson et al., "Two Phase 3 Trials of Dupilumab versus Placebo in Atopic Dermatitis," *New England Journal of Medicine* 375, no. 24 (December 15, 2016): 2335–48, http://www.nejm.org/doi/full/10.1056/NEJMoa1610020.

10. Writing Group for the Women's Health Initiative Investigators, "Risks and Benefits of Estrogen plus Progestin in Healthy Postmenopausal Women," *Journal of the American Medical Association* 288, no. 3 (July 17, 2002): 321–33, http://jama.jamanetwork.com/article.aspx?articleid=195120. The paper also had to adjust for the fact that different women were followed for different numbers of years.

11. For example, a study of 2,763 postmenopausal women with coronary disease concluded that "there were no significant differences between [hormone replacement and placebo] groups in the primary outcome [heart attacks] or in any of the secondary cardiovascular outcomes." Stephen Hulley et al., "Randomized Trial of Estrogen Plus Progestin for Secondary Prevention of Coronary Heart Disease in Postmenopausal Women," *Journal of the American Medical Association* 280, no. 7 (August 19, 1998): 605–13, https://jamanetwork.com/journals/jama/fullarticle/187879.

12. See e.g. the entertaining talk by Ben Goldacre, "Battling Bad Science," filmed 2011, TED talk, 14:13, https://www.ted.com/talks/ben_goldacre_battling_bad_science.

13. See Spencer Jakab, "Is Your Stockpicker Lucky or Good?," *Wall Street Journal*, November 24, 2017, https://www.wsj.com/articles/is-your-stockpicker-lucky-or-good-1511519400. The story reports on a paper by James White, Jeffrey Rosenbluth, and Victor Haghani, "What's Past Is Not Prologue," posted online on September 12, 2017, at https://papers.ssrn.com/sol3/papers.cfm?abstract_id=3034686 (thanks to Paul Rossi for the link).

14. One way to think of this is as follows: if we flip each coin n times, and let X be the number of heads on the 60 percent coin *minus* the number of heads on the 50 percent coin, then we will guess correctly whenever X is greater than zero. Here, X has a mean of $n \times 0.1$ and a variance of $n \times 0.49$. Using the normal approximation with the continuity correction, $Prob(X > 0)$ is approximately the probability that a normal random variable with a mean of $n \times 0.1$ and a variance of $n \times 0.49$ will be larger than 0.5. This probability equals 0.9503843 when $n = 143$, but only 0.9497462 when $n = 142$. Alternatively, Appendix A of the paper by White et al. (see note 13) uses double binomial sums and obtains the same answer: 143.

15. See *Climate Change 2014: Synthesis Report* (Geneva: Intergovernmental Panel on Climate Change, 2015), 2, http://www.ipcc.ch/report/ar5/syr/.

16. See "Climate 101," Climate Reality Project, https://www.climaterealityproject.org/climate-101.

17. See "Understand: Climate Change," US Global Change Research Program, https://www.globalchange.gov/climate-change.

18. Donald J. Trump (@realDonaldTrump), Twitter, November 6, 2012, 2:15 p.m., https://twitter.com/realdonaldtrump/status/265895292191248385.

19. James M. Inhofe, *The Greatest Hoax: How the Global Warming Conspiracy Threatens Your Future* (Washington, DC: WND Books, 2012). Related statements by Inhofe can be found at "James M. Inhofe," DeSmog: Clearing the PR Pollution That Clouds Climate Science, https://www.desmogblog.com/james-inhofe.

20. *The Great Global Warming Swindle*, directed and written by Martin Durkin, aired March 8, 2007, on Channel 4 (UK). See also Al Webb, "Global Warming Labeled a 'Scam,'" *Washington Times*, March 6, 2007, https://web.archive.org/web/20070308093308/http://www.washtimes.com/world/20070306-122226-6282r.htm.

21. For a description, see "Global Climate Change: Vital Signs of the Planet," NASA, https://climate.nasa.gov/vital-signs/global-temperature/. The raw data can be downloaded from https://climate.nasa.gov/system/internal_resources/details/original/647_Global_Temperature_Data_File.txt.

22. For example, a t-test for the difference of means in those two 37-year periods gives a p-value less than 2.2×10^{-16}, and a linear regression of annual temperature versus year gives a regression coefficient of 0.007152 degrees Celsius per year with a p-value less than 2.2×10^{-16}. Both are extremely statistically significant.

23. See e.g. the discussion at BackgammonMasters.com (http://www.backgammonmasters.com/the-growing-popularity-of-backgammon.shtml). For a list of backgammon websites, see Tom Keith, "Backgammon Play Sites," Backgammon Galore, http://www.bkgm.com/servers.html.

24. See e.g. the many threads in the discussion forum at Backgammon Galore (http://www.bkgm.com/rgb/rgb.cgi?menu+computerdice).

25. As an example, the makers of Backgammon NJ devote a page on their site (http://www.njsoftware.com/note.html) to reviews and other arguments that their game is "honest."

26. See Jeff Rollason, "Backgammon Programs Cheat: Urban Myth??," *AI Factory Newsletter* (summer 2010), http://www.aifactory.co.uk/newsletter/2010_01_backgammon_myth.htm (thanks to my brother Alan for the link). Rollason describes a trial he ran, simulating 200,000 users playing twice a day for five days. He writes that he observed an average of roughly 153 doubles per user. In fact, the average was 15,286,212 / 200,000 = 76.43 doubles per user, so I have used the correct figures in my calculations herein.

27. This follows from the R command "pbinom(76*.6, 76, 0.5, lower.tail=FALSE)", which gives an answer of 0.04232305.

28. For one of many such examples, see "Testimonials," Siskiyou Vital Medicine, https://www.siskiyouvitalmedicine.com/client-testimonials/.

29. See the details of her story at http://www.ariplex.com/ama/amamiche.htm. (Warning: includes graphic images.)

30. David H. Gorski (posting as Orac), "A Different Kind of Alternative Medicine 'Testimonial,'" *Respectful Insolence*, November 8, 2006, http://scienceblogs.com/insolence/2006/11/08/a-different-kind-of-testimonial/.

31. For example, the American Cancer Society claims five-year survival rates of nearly 100 percent for stage 0 and stage 1 breast cancer, 93 percent for stage 2, 72 percent for stage 3, and even 22 percent for stage 4; see "Breast Cancer Survival Rates," American Cancer Society, https://www.cancer.org/cancer/breast-cancer/understanding-a-breast-cancer-diagnosis/breast-cancer-survival-rates.html.

Chapter 13: Repeated Luck

1. The friend was my old Harvard student colleague and roommate Marc Goldman.

2. The corresponding commands in R for each of the five gamblers are, respectively: "pbinom(15, 30, 18/38, lower.tail=FALSE)", "pbinom(54, 100, 18/38, lower.tail =FALSE)", "pbinom(28, 50, 18/38, lower.tail=FALSE)", "pbinom(12, 20, 18/38, lower.tail=FALSE)", and "pbinom(1000, 2000, 18/38, lower.tail=FALSE)", leading, respectively, to the p-values 0.3181193, 0.07668926, 0.0863184, 0.08747805, and 0.009815736.

3. To the Empire Financial Group on April 25, 2006, at the Hyatt Regency Maui Resort & Spa, in Hawaii.

4. The cartoon, entitled "Significant," can be found at https://xkcd.com/882/.

5. A.J. Wakefield et al., "Ileal-Lymphoid-Nodular Hyperplasia, Non-specific Colitis, and Pervasive Developmental Disorder in Children," *Lancet* 351, no. 9103 (February 28, 1998): 637–41, http://www.thelancet.com/journals/lancet/article/PIIS0140-6736(97)11096-0/abstract. The paper was later retracted.

6. See "Confirmed Cases of Measles, Mumps, and Rubella, 1996–2013," Public Health England, http://webarchive.nationalarchives.gov.uk/20140505192926/http://www.hpa.org.uk/web/HPAweb&HPAwebStandard/HPAweb_C/1195733833790.

7. See e.g. Kreesten Meldgaard Madsen et al., "A Population-Based Study of Measles, Mumps, and Rubella Vaccination and Autism," *New England Journal of Medicine* 347, no. 19 (November 7, 2002): 1477–82, http://www.nejm.org/doi/full/10.1056/NEJMoa021134.

8. See Editors of *The Lancet*, "Retraction—Ileal-Lymphoid-Nodular Hyperplasia, Non-specific Colitis, and Pervasive Developmental Disorder in Children," *Lancet* 375, no. 9713 (February 6, 2010), http://www.thelancet.com/journals/lancet/article/PIIS0140-6736(10)60175-4/fulltext.

9. See e.g. Sarah Boseley, "Andrew Wakefield Found 'Irresponsible' by GMC over MMR Vaccine Scare," *Guardian* (London), January 28, 2010, https://www.theguardian.com/society/2010/jan/28/andrew-wakefield-mmr-vaccine and James Meikle and Sarah Boseley, "MMR Row Doctor Andrew Wakefield Struck Off Register," *Guardian* (London), May 24, 2010, https://www.theguardian.com/society/2010/may/24/mmr-doctor-andrew-wakefield-struck-off.

10. Fiona Godlee, Jane Smith, and Harvey Marcovitch, "Wakefield's Article Linking MMR Vaccine and Autism Was Fraudulent," *British Medical Journal* 342 (January 6, 2011), http://www.bmj.com/content/342/bmj.c7452.full.

11. See Lyn Redwood, "Why Aren't I Surprised that the Media Got It Wrong AGAIN?," SafeMinds, October 5, 2015, http://www.safeminds.org/blog/2015/10/05/why-arent-i-surprised-that-the-media-got-it-wrong-again/.

12. Bharathi S. Gadad et al., "Administration of Thimerosal-Containing Vaccines to Infant Rhesus Macaques Does Not Result in Autism-like Behavior or Neuropathology," *Proceedings of the National Academy of Science* 112, no. 40 (October 6, 2015):12498–503, http://www.pnas.org/content/112/40/12498.full.

13. Jessica R. Biesiekierski et al., "Gluten Causes Gastrointestinal Symptoms in

Subjects without Celiac Disease: A Double-Blind Randomized Placebo-Controlled Trial," *American Journal of Gastroenterology* 106, no. 3 (March 2011): 508–14. A summary is available at https://www.ncbi.nlm.nih.gov/pubmed/21224837.

14. Jessica R. Biesiekierski et al., "No Effects of Gluten in Patients with Self-Reported Non-celiac Gluten Sensitivity after Dietary Reduction of Fermentable, Poorly Absorbed, Short-Chain Carbohydrates," *Gastroenterology* 145, no. 2 (August 2013): 320–28, http://www.gastrojournal.org/article/S0016-5085(13)00702-6/fulltext.

15. See e.g. Rebecca Davis, "The Doctor Who Championed Hand-Washing and Briefly Saved Lives," *Shots: Health News from NPR,* January 12, 2015, http://www.npr.org/sections/health-shots/2015/01/12/375663920/the-doctor-who-championed-hand-washing-and-saved-women-s-lives.

16. Pasteur's patent for this process can be viewed at https://www.google.com/patents/US135245.

17. See e.g. the lengthy discussion in John Farley and Gerald L. Geison, "Science, Politics and Spontaneous Generation in Nineteenth-Century France: The Pasteur-Pouchet Debate," *Bulletin of the History of Medicine* 48, no. 2 (Summer 1974): 161–98.

18. According to Wikipedia's article on the French franc (https://en.wikipedia.org/wiki/French_franc#Latin_Monetary_Union), in 1865 it was equivalent to about 0.29 grams of gold. And according to an online source (http://www.goldpriceoz.com/gold-price-us/), at the time of this writing, gold traded at US\$1209.80 per troy ounce. Furthermore, one gram is equal to 0.032151 troy ounces. This gives a value for Pasteur's prize of 2500 $\times$ 0.29 $\times$ 0.032151 $\times$ 1209.8 = US\$28,199.80.

19. See e.g. "Louis Pasteur," Biography.com, https://www.biography.com/people/louis-pasteur-9434402 and Mihai Andrei, "5 Things Louis Pasteur Did to Change the World," ZME Science, May 11, 2015, https://www.zmescience.com/other/feature-post/louis-pasteur-changed-world/.

20. See e.g. a widely cited article: John P.A. Ioannidis, "Why Most Published Research Findings Are False," *PLoS Medicine* 2, no. 8 (August 2005): e124, http://journals.plos.org/plosmedicine/article?id=10.1371/journal.pmed.0020124.

21. See Open Science Collaboration, "Estimating the Reproducibility of Psychological Science," *Science* 349, no. 6251 (August 28, 2015), http://science.sciencemag.org/content/349/6251/aac4716. See also Ian Sample, "Study Delivers Bleak Verdict on Validity of Psychology Experiment Results," *Guardian* (London), August 27, 2015, https://www.theguardian.com/science/2015/aug/27/study-delivers-bleak-verdict-on-validity-of-psychology-experiment-results.

22. Ashley Marcin, "Yellow, Brown, Green, and More: What Does the Color of My Phlegm Mean?," Healthline, http://www.healthline.com/health/green-phlegm.

23. Family Health Team, "What the Color of Your Snot Really Means," Cleveland Clinic, June 28, 2017, https://health.clevelandclinic.org/2017/06/what-the-color-of-your-snot-really-means/.

24. Robert H. Shmerling, "Don't Judge Your Mucus by Its Color," *Harvard Health Blog,* February 8, 2016, http://www.health.harvard.edu/blog/dont-judge-your-mucus-by-its-color-201602089129.

25. John Turnidge, "Health Check: Does Green Mucus Mean You're Infectious and Need Antibiotics?," *Conversation*, http://theconversation.com/health-check-does -green-mucus-mean-youre-infectious-and-need-antibiotics-63193.

26. See e.g. the amusing article by Christie Aschwanden, "Café or Nay?," *Slate*, July 27, 2011, http://www.slate.com/articles/health_and_science/medical_examiner/ 2011/07/caf_or_nay.html.

27. See e.g. Jeff Donn, "Medical Benefits of Dental Floss Unproven," Associated Press, August 2, 2016, https://apnews.com/f7e66079d9ba4b4985d7af350619a9e3/medical -benefits-dental-floss-unproven. See also the following meta-analysis articles: C.E. Berchier et al., "The Efficacy of Dental Floss in Addition to a Toothbrush on Plaque and Parameters of Gingival Inflammation: A Systematic Review," *International Journal of Dental Hygiene* 6, no. 4 (November 2008): 265–79, https:// www.ncbi.nlm.nih.gov/pubmed/19138178; Dario Sambunjak et al., "Flossing for the Management of Periodontal Diseases and Dental Caries in Adults," *Cochrane Database of Systematic Reviews* 2011, no. 12 (December 7, 2011), https://www.ncbi .nlm.nih.gov/pubmed/22161438; and Sonja Sälzer et al., "Efficacy of Inter-dental Mechanical Plaque Control in Managing Gingivitis—A Meta-Review," *Journal of Clinical Periodontology* 42, no. S16 (April 2015): S92–S105, https://www.ncbi.nlm .nih.gov/pubmed/25581718.

28. The results were written up in Raphael Silberzahn et al., "Many Analysts, One Dataset: Making Transparent How Variations in Analytical Choices Affect Results," preprint, submitted September 21, 2017, https://psyarxiv.com/qkwst/.

29. See the original article: Paolo Zamboni et al., "Chronic Cerebrospinal Venous Insufficiency in Patients with Multiple Sclerosis," *Journal of Neurology, Neurosurgery & Psychiatry* 80, no. 4 (April 2009), http://jnnp.bmj.com/content/80/4/392.

30. See e.g. Kelly Crowe, "'Scientific Quackery': UBC Study Says It's Debunked Controversial MS Procedure," CBC News, March 8, 2017, http://www.cbc.ca/news/ health/multiple-sclerosis-liberation-therapy-clinical-trial-1.4014494.

31. See e.g. Ed Yong, "Psychology's Replication Crisis Can't Be Wished Away," *Atlantic*, March 4, 2016, https://www.theatlantic.com/science/archive/2016/03/psychologys -replication-crisis-cant-be-wished-away/472272/.

32. The call was from a producer of the CBC Radio program *The Current*, on October 29, 2010. He was preparing to interview Dr. Ioannidis after reading an article about invalid medical studies (David H. Freeman, "Lies, Damned Lies, and Medical Science," *Atlantic*, November 2010, https://www.theatlantic.com/magazine/ archive/2010/11/lies-damned-lies-and-medical-science/308269/), and, uncertain as to how to proceed, he emailed me with a request "to chat with you. I need to clarify some of my own thinking on this."

33. Brian A. Nosek, Jeffrey R. Spies, and Matt Motyl, "Scientific Utopia: II. Restructuring Incentives and Practices to Promote Truth over Publishability," *Perspectives on Psychological Science* 7, no. 6 (November 2012): 615–31, http:// journals.sagepub.com/doi/full/10.1177/1745691612459058.

34. See Christie Aschwanden, "Science Isn't Broken," *FiveThirtyEight*, August 19, 2015, https://fivethirtyeight.com/features/science-isnt-broken/.

35. See e.g. the 73-author paper Daniel J. Benjamin et al., "Redefine Statistical Significance," preprint, submitted July 22, 2017, https://osf.io/preprints/psyarxiv/mky9j/. See also the follow-up discussion: Dalmeet Singh Chawla, "Big Names in Statistics Want to Shake Up Much-Maligned P Value," *Nature* 548, no. 7665 (August 3, 2017), http://www.nature.com/news/big-names-in-statistics-want-to-shake-up-much-maligned-p-value-1.22375.

36. See e.g. Jonathan W. Schooler, "Metascience Could Rescue the 'Replication Crisis,'" *Nature* 515, no. 7525 (November 6, 2014), http://www.nature.com/news/metascience-could-rescue-the-replication-crisis-1.16275.

37. An editorial on page 1 of *Basic and Applied Social Psychology* (BASP) explicitly said that "BASP is banning NHSTP," meaning the "null hypothesis significance testing procedure"—that is, the use of p-values. David Traflmow and Michael Marks, "Editorial," *Basic and Applied Social Psychology* 37 (2015), http://www.tandfonline.com/doi/abs/10.1080/01973533.2015.1012991?journalCode=hbas20.

38. See e.g. Chris Woolston, "Psychology Journal Bans *P* Values," *Nature* 519, no. 7541 (March 5, 2015), http://www.nature.com/news/psychology-journal-bans-p-values-1.17001. See also Ronald L. Wasserstein, "ASA Comment on a Journal's Ban on Null Hypothesis Statistical Testing," ASA Community, February 26, 2015, http://community.amstat.org/blogs/ronald-wasserstein/2015/02/26/asa-comment-on-a-journals-ban-on-null-hypothesis-statistical-testing and Daniel Lakens, "So You Banned P-values, How's That Working Out for You?," *20% Statistician*, February 10, 2016, http://daniellakens.blogspot.ca/2016/02/so-you-banned-p-values-hows-that.html.

39. Jane Qiu, "Venous Abnormalities and Multiple Sclerosis: Another Breakthrough Claim?," *Lancet: Neurology* 9, no. 5 (May 2010): 464–65, http://www.thelancet.com/journals/laneur/article/PIIS1474-4422(10)70098-3/fulltext.

40. See e.g. "Editorial: Reality Check on Reproducibility," *Nature* 533, no. 7604 (May 26, 2016), https://www.nature.com/news/reality-check-on-reproducibility-1.19961.

Chapter 14: Lottery Luck

1. See e.g. Grant Rodgers, "Guilty Verdict in Hot Lotto Scam, but Game Safe, Official Says," *Des Moines Register*, July 20, 2015, http://www.desmoinesregister.com/story/news/crime-and-courts/2015/07/20/hot-lotto-verdict/30411901/.

2. See e.g. Grant Rodgers, "Tipton Brothers Plead Guilty in Iowa Lottery Rigging Scandal," *Des Moines Register*, June 29, 2017, http://www.desmoinesregister.com/story/news/crime-and-courts/2017/06/29/tipton-pleads-guilty-iowa-lottery-rigging-scandal/438039001/ and Jason Clayworth, "'I Certainly Regret' Rigging Iowa Lottery, Says Cheat Who Gets 25 Years," *Des Moines Register*, August 22, 2017, http://www.desmoinesregister.com/story/news/investigations/2017/08/22/iowa-lottery-cheat-sentenced-25-years/566642001/.

3. See e.g. Harriet Alexander, "World's Largest Lottery Winners Come Forward to Claim Share of $1.58bn Jackpot," *Telegraph* (London), February 17, 2016, http://www.telegraph.co.uk/news/worldnews/northamerica/usa/12162274/Worlds-largest-lottery-winners-come-forward-to-claim-share-of-1.58bn-jackpot.html.

4. For example, the City of Buffalo had total revenues of $1.439 billion in 2016–17; see City of Buffalo, *Fiscal Year 2016–2017: Adopted Budget Detail*, https://www.ci.buffalo.ny.us/Mayor/Home/Leadership/FiscalReporting/Archived_Budgets/20162017AdoptedBudget.

5. According to data from the National Weather Service (http://www.lightningsafety.noaa.gov/fatalities.shtml), there are about 31 lightning fatalities per year in the US, out of a population of about 320 million, corresponding to one American in just over ten million—or about 28 times more likely than one chance in 292 million.

6. The United States had 45 presidents in its first 241 years, or about one president every 5.4 years. The average age at inauguration is 55.0 years (see http://www.presidenstory.com/stat_age.php). Hence, there are about 55.0/5.4, or just over ten future presidents, currently alive. So a randomly chosen person has ten chances in 320 million, or one chance in 32 million, of being a future president. This is just over nine times more likely than one chance in 292 million.

7. As discussed on page 83 of my previous book *Struck by Lightning*, each drive across town has approximately one chance in seven million of resulting in death. This is just under 42 times more likely than one chance in 292 million.

8. There are 62.9 births per year per 1,000 American women aged 15 to 44 years, according to statistics from the National Center for Health Statistics at the Centers for Disease Control and Prevention (https://www.cdc.gov/nchs/fastats/births.htm). So the probability that a randomly chosen woman of this age will give birth in the next second is 62.9 / 1000 / 365 / 24 / 60 / 60, or about one chance in 501 million. In 1.7 seconds, therefore, the probability is about one chance in 501 / 1.7 million, roughly equal to one chance in 292 million.

9. On average, you will win once every 292 million weeks, which corresponds to 292 million / 52 = 5.6 million years.

10. For Global News in Ontario, on the day (Wednesday, October 26, 2005) that a large $54 million Lotto 6/49 jackpot was up for grabs.

11. See the website Lotto 6/49 Stats (http://lotto649stats.com/).

12. He is Richard Lustig, author of the book *Learn to Increase Your Chances of Winning the Lottery*. See e.g. Josh K. Elliott, "How to Boost Your Horrible Odds of Winning the Powerball," CTV News, January 13, 2016, http://www.ctvnews.ca/canada/how-to-boost-your-horrible-odds-of-winning-the-powerball-1.2735726.

13. See e.g. "Lotto 6/49 Ticket Worth $1.6M Sold in Windsor," CTV News Windsor, April 7, 2016, http://windsor.ctvnews.ca/lotto-6-49-ticket-worth-1-6m-sold-in-windsor-1.2849566.

14. For more details about the lottery retailer scandal, see Jeffrey S. Rosenthal, "Statistics and the Ontario Lottery Retailer Scandal," *Chance* 27, no. 1 (February 2014), available at http://probability.ca/lotteryscandal/.

15. The Ontario Lottery and Gaming Corporation, which runs the lottery, soon conducted its own survey and got a factor of 1.95. Corporate Research Associates Inc. later conducted a more detailed survey in Atlantic Canada and obtained a factor of 1.52, virtually identical to the *Fifth Estate* figure of 1.5.
16. This can be computed in R using the command "ppois(199, 57, lower.tail=FALSE)," which gives an answer of 4.653685e-49.
17. The full broadcast is available at http://www.cbc.ca/fifth/episodes/from-the -archives/luck-of-the-draw.
18. See e.g. Rob Ferguson and Curtis Rush, "Province to Probe the Windfalls of Lottery Retailers," *Toronto Star*, October 26, 2006, archived online at http://probability.ca/ jeff/writing/starlott.html.
19. See e.g. the summaries archived at http://probability.ca/sbl/OLG-FAQ.html#10.
20. See e.g. Ontario, *Legislative Assembly of Ontario—Oral Questions*, October 25, 2006, http://www.ontla.on.ca/house-proceedings/transcripts/files_html/2006-10 -25_L113A.htm#P232_25936.
21. "A Game of Trust," Ombudsman Ontario, March 26, 2007, https://www.ombudsman .on.ca/resources/reports-and-case-summaries/reports-on-investigations/2007/a -game-of-trust.
22. For more details, see again my article "Statistics and the Ontario Lottery Retailer Scandal" cited above, and the many references therein.
23. For more details, see Chris Hansen, "How Lucky Can You Get?," *Hansen Files on Dateline*, NBC News, http://www.nbcnews.com/id/38778571/ns/dateline_nbc -the_hansen_files_with_chris_hansen/#.V-k5IiXPHS0.
24. These events took place in Oldham, just outside of Manchester. The customer was Maureen Holt, a 77-year-old great-grandmother. The clerk was Farrakh Nizzar, who was to be deported back to Pakistan after serving his sentence. For more details, see "Lottery Gran on Conman: 'Everyone Calls Him Lucky but He Wasn't Very Lucky This Time,'" *Manchester Evening News*, August 1, 2012, http://www .manchestereveningnews.co.uk/news/greater-manchester-news/lottery-gran-on -conman-everyone-calls-692165 and "New Lottery 'Win' Alert after Shopkeeper Tried to Con Great-Gran from Oldham out of £1m," *Manchester Evening News*, August 27, 2012, http://www.manchestereveningnews.co.uk/news/greater -manchester-news/new-lottery-win-alert-after-693924.

Chapter 15: Lucky Me
1. The Toronto Police Service Fraud Investigators Conference, on December 10–14, 2007.
2. J. Kelly Nestruck, "The Deal Breaker: If You're a Guest on Howie Mandel's Show, You Should Bring Jeffrey Rosenthal—Not Your Dad," *National Post*, May 30, 2006, archived at http://www.probability.ca/lotteryscandal/ref/2006-05-30-post.txt.
3. For more about this, see e.g. Jeffrey Rosenthal, "Improv and Music: An Unusual Duo," Theatresports Toronto newsletter, November 2001, http://probability.ca/jeff/ writing/improvmusic.html.

4. For a performance of the Warren Graves play *The Mumberley Inheritance* at the Scarborough Village Theatre in June 2015, directed by Mike Ranieri. See the show poster at http://probability.ca/jeff/MI/poster.jpg; for information, see the show's Facebook page (https://www.facebook.com/mumberley/); for reviews, see Danny Gaisin, "'The Mumberley Inheritance'; v- 2.0, ... Giggle, Giggle, Giggle!," *Ontario Arts Review*, June 5, 2015, https://ontarioartsreview.ca/2015/06/05/the -mumberley-inheritance-v-2-0-giggle-giggle-giggle/ and Maria Tzavaras, "Cast's Comedic Ability Highlighted in The Mumberley Inheritance," *Scarborough Mirror*, June 12, 2015, https://www.insidetoronto.com/news-story/5675231-cast -s-comedic-ability-highlighted-in-the-mumberley-inheritance/.

5. At the Richmond Hill Centre for the Performing Arts. Information on the show appears at http://sa1.seatadvisor.com/sabo/servlets/EventInfo?eventId=1159161 and https://www.facebook.com/events/859257057564334/. I provided musical accompaniment for the well-known improv troupe Not to Be Repeated (an entry in IMDB for their television show, *This Sitcom Is . . . Not to Be Repeated*, appears at http://www.imdb.com/title/tt0305127/).

6. Someone kindly sent me a photo of the incident, which I posted on my website at http://probability.ca/jeff/images/juggling_shapiro.jpg.

7. See e.g. Jennifer Yang, "Numbers Don't Always Tell the Whole Story," *Toronto Star*, January 30, 2010, https://www.thestar.com/news/gta/2010/01/30/numbers_dont_ always_tell_the_whole_story.html or the article "Not So Rare for Rarities to Occur in Waves: Professor," *Metro* (Toronto), January 29, 2010, http://www.metronews.ca/ news/toronto/2010/01/29/not-so-rare-for-rarities-to-occur-in-waves-professor.html.

Chapter 16: Lucky Sports

1. See Kelly Phillips Erb, "Warren Buffett Offers $1 Billion for Perfect March Madness Bracket," Forbes, January 21, 2014, https://www.forbes.com/sites/kellyphillipserb/ 2014/01/21/warren-buffett-offers-1-billion-for-perfect-march-madness-bracket/ #72862857100b.

2. See Jeffrey Rosenthal, "Rosenthal: A Statistical Ranking of NCAA Basketball Teams," TSN, March 18, 2013, http://www2.tsn.ca/ncaa/story/?id=418503.

3. The boxscore of the Oklahoma State–Oregon game can be found at http://www .sports-reference.com/cbb/boxscores/2013-03-21-oklahoma-state.html.

4. The boxscore of the Harvard–New Mexico game can be found at http://www.ncaa .com/game/basketball-men/d1/2013/03/21/harvard-new-mexico.

5. See "NCAA Basketball Tournament History: Harvard Crimson," ESPN, http:// www.espn.com/mens-college-basketball/tournament/history/_/team1/6128.

6. See Peter Kim, "Toronto Blue Jays Have 88.52% Chance of Making the Playoffs: Stats Professor," Global News, September 22, 2015, http://globalnews.ca/ news/2235467/toronto-blue-jays-have-88-52-chance-of-making-the-playoffs-stats -professor/.

7. The boxscore of the Blue Jays–Yankees game can be found at http://www.baseball -reference.com/boxes/TOR/TOR201509220.shtml.

8. The final standings for the 2015 season can be found at http://www.baseball -reference.com/leagues/MLB/2015-standings.shtml#all_standings_E.

9. NHL standings on April 11, 2006, can be found at http://www.hockey-reference .com/boxscores/index.cgi?month=4&day=11&year=2006.

10. Mike Strobel, "According to the School of Biased Observation, It's Fated that the Leafs Are Going to the Cup This Year," *Toronto Sun*, April 13, 2006, archived at http://probability.ca/lotteryscandal/ref/2006-04-13-sun.txt. See also the email I sent Strobel, archived at http://probability.ca/lotteryscandal/ref/NHLmesg.txt.

11. Final NHL standings for the 2005–06 season can be found on the league's website at https://www.nhl.com/standings/2005.

12. See "Career Leaders & Records for Batting Average," Baseball Reference, https:// www.baseball-reference.com/leaders/batting_avg_career.shtml.

13. See "30+ Game Hitting Streaks," *Baseball Almanac*, http://www.baseball-almanac .com/feats/feats-streak.shtml.

14. See e.g. DiMaggio's official baseball statistics at http://m.mlb.com/player/113376/ joe-dimaggio.

15. I computed this in R with the command "pbinom(0, 4, 0.3246, lowertail=FALSE)$\hat{5}$6," which works out to about one chance in 472,118. This simplified calculation assumes exactly four at-bats in each game, and that the different games are independent. Correcting these assumptions would be possible, but not simple. Anyway, DiMaggio was fairly consistent, with batting averages between .290 and .357 except for his final season (1951, when he batted .263) and his best year (1939, when he batted .381), suggesting that these assumptions aren't so unreasonable.

16. For this, I used a simple Monte Carlo simulation. I randomly simulated a sequence of 1,736 games, each having at least one hit with probability 0.7919133. I then computed the longest hitting streak within the sequence. Repeating this simulation 100,000 times, I found that the absolute longest hitting streak was 75, but the *average* longest hitting streak was 27.21943, and furthermore, the fraction of simulations with a streak of 56 games or longer was 0.00067, or about one chance in 1,500. My simple R computer program is available for inspection at http://probability.ca/kow/ Rdimag.txt. It might be possible to compute this probability analytically using the "inclusion-exclusion principle" formula, though the calculations seem messy; enterprising readers are invited to attempt that computation and let me know.

17. See e.g. Eric Fisher, "MLBAM's Beat the Streak Chases History," *Sports Business Journal*, May 16, 2016, https://www.sportsbusinessdaily.com/Journal/ Issues/2016/05/16/Leagues-and-Governing-Bodies/MLBAM-beat-the-streak.aspx. To enter the Beat the Streak contest, visit http://mlb.mlb.com/mlb/fantasy/bts/.

18. This is what DiMaggio supposedly "confided to a teammate." See e.g. "1941: Joe DiMaggio Ends 56-Game Hitting Streak," *This Day in History*, History.com, July 17, http://www.history.com/this-day-in-history/joe-dimaggio-ends-56-game-hitting -streak and Dave Whitehorn, "20 Fun Facts about Joe DiMaggio's 56-Game Hit Streak," *Newsday*, May 11, 2016, https://www.newsday.com/sports/baseball/ yankees/joe-dimaggio-s-56-game-hit-streak-20-fun-facts-1.3028286.

19. See e.g. Christopher Dabe, "New Orleans Saints LB Stephone Anthony Named to PFWA All-Rookie Team," *Times-Picayune* (New Orleans), January 19, 2016, http://www.nola.com/saints/index.ssf/2016/01/new_orleans_saints_stephone_an_1.html; Sam Robinson, "Stephone Anthony's Disappointing Second Season to End on IR," *Fanrag Sports Network*, December 20, 2016, https://www.fanragsports.com/news/stephone-anthonys-disappointing-second-season-end-ir/; and Marc Sessler, "Dolphins Acquire LB Stephone Anthony from Saints," *Around the NFL*, NFL.com, September 19, 2017, http://www.nfl.com/news/story/0ap3000000848297/article/dolphins-acquire-lb-stephone-anthony-from-saints.

20. For a nice illustration of this combination in different professional sports games, see e.g. Vox, "Why Underdogs Do Better in Hockey than Basketball," YouTube, uploaded June 5, 2017, https://www.youtube.com/watch?v=HNlgISa9Giw, based on Michael J. Mauboussin, *The Success Equation* (Boston: Harvard Business Review Press, 2012).

Chapter 17: Lucky Polls

1. See e.g. Adam Shergold, "The Man You Can Count On: The Poker-Playing Numbers Expert Who Predicted Presidential Election Outcomes with Incredible Accuracy," *Daily Mail* (London), November 8, 2012, http://www.dailymail.co.uk/news/article-2229790/US-Election-2012-Statistician-Nate-Silver-correctly-predicts-50-states.html.

2. See e.g. the pre-referendum poll summaries reported in Charlie Cooper, "EU Referendum: Final Polls Show Remain with Edge over Brexit," *Independent* (London), June 23, 2016, http://www.independent.co.uk/news/uk/politics/eu-referendum-poll-brexit-remain-vote-leave-live-latest-who-will-win-results-populus-a7097261.html; "Brexit Poll Tracker," *Financial Times* (London), https://ig.ft.com/sites/brexit-polling/; and "EU Referendum Poll of Polls," *What UK Thinks*, https://whatukthinks.org/eu/opinion-polls/poll-of-polls/.

3. Patrick Sturgis et al., *Report of the Inquiry into the 2015 British General Election Opinion Polls* (London: Market Research Society and British Polling Council, 2016), http://eprints.ncrm.ac.uk/3789/.

4. See David Cowling, "Election 2015: How the Opinion Polls Got It Wrong," BBC News, May 17, 2015, http://www.bbc.com/news/uk-politics-32751993.

5. See Tom Clark, "New Research Suggests Why General Election Polls Were So Inaccurate," *Guardian* (London), November 13, 2015, https://www.theguardian.com/politics/2015/nov/13/new-research-general-election-polls-inaccurate.

6. See Anthony Wells, "Election 2015 Polling: A Brief Post Mortem," YouGov, May 8, 2015, https://yougov.co.uk/news/2015/05/08/general-election-opinion-polls-brief-post-mortem/ and Ben Lauderdale, "What We Got Wrong in Our 2015 U.K. General Election Model," *FiveThirtyEight*, May 8, 2015, https://fivethirtyeight.com/features/what-we-got-wrong-in-our-2015-uk-general-election-model/.

7. See e.g. the discussion in Tim Harford, "Big Data: A Big Mistake?," *Significance* 11, no. 5 (December 2014): 14–19, http://onlinelibrary.wiley.com/doi/10.1111/j.1740-9713.2014.00778.x/full and Dennis DeTurck, "Case Study I: The 1936 *Literary*

Digest Poll," https://www.math.upenn.edu/~deturck/m170/wk4/lecture/case1.html. Note that the 3,000 sample size was actually for Gallup's smaller poll, which attempted to predict the results of the *Literary Digest* poll; see P. Squire's article below (note 10).

8. The *Literary Digest* was taken over by *Time* magazine and ceased to exist as a separate publication on May 23, 1938; see the announcement at http://content.time .com/time/magazine/article/0,9171,882981,00.html. The election had been held on November 3, 1936, which was one year, six months, and 20 days earlier.

9. See the bottom graph at "Gallup Presidential Election Trial-Heat Trends, 1936–2008," Gallup, http://www.gallup.com/poll/110548/gallup-presidential -election-trial-heat-trends.aspx.

10. See e.g. Peverill Squire, "Why the 1936 *Literary Digest* Poll Failed," *Public Opinion Quarterly* 52, no. 1 (spring 1988): 125–33.

11. See e.g. David Lauter, "One Last Look at the Polls: Hillary Clinton's Lead Is Holding Steady," *Los Angeles Times*, November 8, 2016, http://www.latimes.com/nation /politics/trailguide/la-na-election-day-2016-a-last-look-at-the-polls-clinton-lead -1478618744-htmlstory.html.

12. By about 48.2 percent to 46.1 percent; see e.g. Gregory Krieg, "It's Official: Clinton Swamps Trump in Popular Vote," CNN, December 22, 2016, http://www.cnn.com/ 2016/12/21/politics/donald-trump-hillary-clinton-popular-vote-final-count/.

13. See e.g. the summary graphic at https://www.dailywire.com/sites/default/files/ uploads/2016/11/rcp_general_election_4_11.7.2016_0.jpg. For a much deeper look at pre-election poll media interpretations, see e.g. Nate Silver, "The Real Story of 2016," *FiveThirtyEight*, January 19, 2017, http://fivethirtyeight.com/features/the -real-story-of-2016/.

14. See "Canada Not Immune to 'Hate Wave': CNN Commentator Van Jones," *Globe and Mail* Video, November 23, 2016, https://www.theglobeandmail.com/news/ news-video/video-canada-not-immune-to-hate-wave-cnn-commentator-van -jones/article33004444/.

15. The final election count had been 48.2 percent for Clinton, 46.1 percent for Trump, and 5.7 percent for other candidates. Suppose a polling company had tried to phone 10,000 people who were perfectly representative: in other words, there would be 4,820 Clinton supporters, 4,610 Trump supporters, and 570 supporting others. If the response rate was 10 percent for Clinton and for "Other" supporters, and 9.6 percent for Trump supporters, then the polling company will get $4,820 \times 10\% \doteq 482$ Clinton responses, $4610 \times 9.6\% \doteq 443$ Trump responses, and $570 \times 10\% \doteq 57$ responses for others. The total number of responses would therefore be $482 + 443 + 57 = 982$, of which Clinton would receive $482 / 982 \doteq 49.1\%$, and Trump would have $443 / 982 \doteq 45.1\%$, giving Clinton a margin of victory of 4 percent.

16. See e.g. David Leip, "2016 Presidential General Election Results [Alabama]," *Dave Leip's Atlas of U.S. Presidential Elections*, https://uselectionatlas.org/RESULTS/state .php?year=2016&fips=1.

17. See e.g. Maegan Vazquez, "Trump Calls Roy Moore to Offer His Endorsement," CNN, December 4, 2017, http://www.cnn.com/2017/12/04/politics/trump-moore-endorsement-twitter/.

18. Respectively, the Emerson College poll at https://www.realclearpolitics.com/docs/Emerson_College_Alabama_Dec_11.pdf; the Fox News poll at http://www.foxnews.com/politics/2017/12/11/fox-news-poll-enthused-democrats-give-jones-lead-over-moore-in-alabama.html; and the Monmouth University poll at https://www.monmouth.edu/polling-institute/reports/MonmouthPoll_AL_121117/. See also the summary of polls by Harry Enten, "Everything You Need to Know about Alabama's Senate Election," *FiveThirtyEight*, December 12, 2017, https://fivethirtyeight.com/features/everything-you-need-to-know-about-alabamas-senate-election/.

19. See Nate Silver, "What the Hell Is Happening with These Alabama Polls?," *FiveThirtyEight*, December 11, 2017, https://fivethirtyeight.com/features/what-the-hell-is-happening-with-these-alabama-polls/. For what it's worth, Silver argued that Moore was polling higher in automated "interactive voice response" (IVR, also known as "robocall") polls than in traditional live-caller polls, and speculated that this might be due to such factors as automated polls not reaching younger voters with cellphones.

20. See Brett LoGiurato (@BrettLoGiurato), Twitter, December 11, 2017, 10:21 a.m., https://twitter.com/BrettLoGiurato/status/940240018005745664.

21. The great Andy Barrie, on CBC Radio's *Metro Morning*, at 6:40 a.m. on December, 22, 2005, during the run-up to the January 2006 Canadian federal election.

Chapter 18: Interlude: Lucky Sayings

1. See "Aphorism," Dictionary.com, http://www.dictionary.com/browse/aphorism.

2. For instance, a 2011 FBI report says that, while there are over 200,000 child abductions in the United States each year, most of them are related to family custody battles, and only about 115 of the reported abductions each year involve strangers abducting children for ransom or to kill or keep them. Out of a total of 74.2 million children in the US, this represents about one child in 575,000. Ashli-Jade Douglas, "Child Abductions: Known Relationships Are the Greater Danger," *FBI Law Enforcement Bulletin*, August 1, 2011, https://leb.fbi.gov/2011/august/crimes-against-children-spotlight-child-abductions-known-relationships-are-the-greater-danger.

3. See "Audentes Fortuna Juvat," Merriam-Webster, https://www.merriam-webster.com/dictionary/audentesfortunajuvat.

4. In the 1986 movie *Star Trek IV: The Voyage Home*. See e.g. "*Star Trek IV: The Voyage Home*—Quotes," IMDB, http://www.imdb.com/title/tt0092007/quotes.

5. Bob Dylan, in the 1975 song "Simple Twist of Fate." See Bob Dylan, "Simple Twist of Fate Lyrics," MetroLyrics, http://www.metrolyrics.com/simple-twist-of-fate-lyrics-bob-dylan.html. The Nobel Prize announcement is at: https://www.nobelprize.org/nobel_prizes/literature/laureates/2016/.

6. See e.g. "Luck of the Irish," *Urban Dictionary*, http://www.urbandictionary.com/define.php?term=luck of the irish.

7. See e.g. "Where Does the Term 'The Luck of the Irish' Come From?," *Irish Central*, August 8, 2017, https://www.irishcentral.com/roots/history/where-does-the-term -the-luck-of-the-irish-come-from, which quotes Professor E.T. O'Donnell of Holy Cross College.

8. See Richard Wiseman, "The Luck Factor," *Skeptical Inquirer* 27, no. 3 (May/June 2003), http://www.richardwiseman.com/resources/The_Luck_Factor.pdf.

9. Credited to John Clarke's *Parœmiologia Anglo-Latina*; see John Simpson and Jennifer Speake, eds., *The Oxford Dictionary of Proverbs* (Oxford: Oxford University Press, 2009), http://www.oxfordreference.com/view/10.1093/ acref/9780199539536.001.0001/acref-9780199539536-e-151.

10. See "Lefty Gomez Quotes," *Baseball Almanac*, http://www.baseball-almanac.com/ quotes/quolgom.shtml.

11. See e.g. Marc Tracy, "Better to Be Lucky than Good? Sometimes It's True," *New York Times*, December 18, 2015, https://www.nytimes.com/2015/12/19/sports/ ncaabasketball/better-to-be-lucky-than-good-sometimes-its-true.html.

12. See Thomas McKelvey Cleaver, "It's Better to Be Lucky than Good," *Defenders of the Philippines*, http://philippine-defenders.lib.wv.us/pdf/bios/gillett_bio.pdf.

13. Available on my website at http://probability.ca/jeff/nonwork/profile.html.

14. See e.g. Claudio, "10 Rags to Riches Millionaire Musicians," *Richest*, January 28, 2014, http://www.therichest.com/rich-list/poorest-list/10-rags-to-riches-millionaire -musicians/.

15. Emily Esfahani Smith, "You'll Never Be Famous—And That's O.K.," *New York Times*, September 4, 2017, https://www.nytimes.com/2017/09/04/opinion/ middlemarch-college-fame.html.

16. Quoted in Daniel A. Vallero, *Paradigms Lost: Learning from Environmental Mistakes, Mishaps, and Misdeeds* (Boston: Butterworth-Heinemann, 2006), 367. See also "Marshall McLuhan Quotes," BrainyQuote.com, https://www.brainy- quote.com/quotes/quotes/m/marshallmc100969.html and Josephine Gross, "We Are All Stewards on Spaceship Earth," EvanCarmichael.com, http://www .evancarmichael.com/library/josephine-gross/We-Are-All-Stewards-on-Spaceship -Earth.html.

Chapter 19: Justice Luck

1. See e.g. Jessica Anderson, "Armed Men Accused of Holding Up a Baltimore County Bar—Where Cops Were Celebrating an Officer's Retirement," *Baltimore Sun*, August 30, 2017, http://www.baltimoresun.com/news/maryland/crime/bs-md-co -retirement-party-robbery-20170830-story.html. See also Kai Reed, "Armed Suspects Rob Pub Full of Police Officers Attending Party," WBAL-TV, September 1, 2017, http://www.wbaltv.com/article/armed-suspects-rob-pub-full-of-police-officers -attending-party/12149896 and Tribune Media Wire, "Men Accused of Trying to Hold Up Bar during Police Retirement Party," WREG-TV, September 1, 2017, http://wreg.com/2017/09/01/men-accused-of-trying-to-hold-up-bar-during -police-retirement-party/.

2. See e.g. Ashifa Kassam, "Prison Escapees Caught at Canadian Escape Room Interactive Game," *Guardian* (London), https://www.theguardian.com/world/2017/oct/05/canada-prison-escapees-caught-escape-room; "Violent Offenders Caught after Escape from Edmonton Institution for Women," CBC News, October 3, 2017, http://www.cbc.ca/news/canada/edmonton/edmonton-institution-women-prisoners-caught-1.4318566; Dustin Coffman and Phil Heidenreich, "2 Women Who Escaped Edmonton Institution for Women Back in Custody: Police," Global News, October 3, 2017, https://globalnews.ca/news/3782025/edmonton-police-searching-for-2-escaped-prisoners/; and Canadian Press, "Police Issue Warning after Two Women Escape from Prison," CTV News, October 3, 2017, http://www.ctvnews.ca/canada/police-issue-warning-after-two-women-escape-from-prison-1.3617229.

3. People v. Collins, 68 Cal.2d 319 (1968). Complete text of the decision can be found at https://scholar.google.com/scholar_case?case=2393563144534950884.

4. For further discussion of this case, see Jeffrey Rosenthal, "Probability, Justice, and the Risk of Wrongful Conviction," *Mathematics Enthusiast* 12 (June 2015): 11–18, available on my website at http://probability.ca/jeff/ftpdir/probjustice.pdf.

5. See e.g. Ray Hill, "Multiple Sudden Infant Deaths—Coincidence or Beyond Coincidence?," *Paediatric and Perinatal Epidemiology* 18, no. 5 (September 2004): 320–26.

6. See Royal Statistical Society, "Royal Statistical Society Concerned by Issues Raised in Sally Clark Case," news release, October 23, 2001, http://www.rss.org.uk/Images/PDF/influencing-change/2017/SallyClarkRSSstatement2001.pdf.

7. Alfred Steinschneider, "Prolonged Apnea and the Sudden Infant Death Syndrome: Clinical and Laboratory Observations," *Pediatrics* 50, no. 4 (October 1972): 646–54.

8. See e.g. George Judson, "Mother Guilty in the Killings of Five Babies," *New York Times*, April 22, 1995.

9. See e.g. Jackie Hong and Jayme Poisson, "Elizabeth Wettlaufer Pleads Guilty to Murdering 8 Seniors," *Toronto Star*, June 1, 2017, https://www.thestar.com/news/canada/2017/06/01/elizabeth-wettlaufer-woodstock-nurse-guilty-murder.html and John Lancaster, "Seeing Red," CBC News, October 6, 2017, http://www.cbc.ca/news2/interactives/sh/TBk79oWhpi/elizabeth-wettlaufer-nurse-senior-deaths/.

10. See e.g. the summary in Ronald Meester, Marieke Collins, Richard Gill, and Michiel van Lambalgen, "On the (Ab)Use of Statistics in the Legal Case against the Nurse Lucia De B.," *Law, Probability and Risk* 5, no. 3–4 (September 2006), available at: arxiv.org/pdf/math/0607340.pdf.

11. A baby girl who died was found to have excess digoxin levels, and a baby boy's coma may have been caused by an overdose of chloral hydrate; see e.g. the discussion at http://www.luciadeb.nl/english/summary.html.

12. See e.g. Marlise Simons, "Court to Rule on Dutch Nurse Accused in 13 Deaths," *New York Times*, October 8, 2002, http://www.nytimes.com/2002/10/08/world/court-to-rule-on-dutch-nurse-accused-in-13-deaths.html.

13. See e.g. Ben Goldacre, "Conviction for Patients' Deaths Does Not Add Up," *Guardian* (London), April 10, 2010, https://www.theguardian.com/commentisfree/2010/apr/10/bad-science-dutch-nurse-case.

14. For example, Meester et al. (cited above) wrote that "the data . . . is used twice: first to identify the suspect, and after that again in the computations of Elffers' probabilities." The authors made numerous "adjustments," and eventually increased the p-value from "1 in 342 million" to 0.022 (or one chance in 45), a p-value that is surely too large to convict.

15. See e.g. Associated Press, "Apology for Nurse Jailed for Murdering Seven Patients," *Independent* (London), April 14, 2010, http://www.independent.co.uk/news/world/europe/apology-for-nurse-jailed-for-murdering-seven-patients-1944577.html.

Chapter 20: Astrological Luck

1. The Wellcome Trust Monitor Survey, in Britain, discussed here: Nick Allum, "Some People Think Astrology Is a Science—Here's Why," Conversation, July 1, 2014, http://theconversation.com/some-people-think-astrology-is-a-science-heres-why-28642.

2. See e.g. this description: "The Chinese Zodiac," China Highlights, https://www.chinahighlights.com/travelguide/chinese-zodiac/.

3. ShaoLan Hsueh, "The Chinese Zodiac, Explained," TED talk, filmed 2016, 6:05, https://www.ted.com/talks/shaolan_the_chinese_zodiac_explained.

4. See Amy Qin, "When Young Chinese Ask, 'What's Your Sign?' They Don't Mean Dragon or Rat," *New York Times*, July 22, 2017, https://www.nytimes.com/2017/07/22/world/asia/china-western-astrology.html.

5. See Nick Allum, "What Makes Some People Think Astrology Is Scientific?," *Science Communication* 33, no. 3 (September 2011): 341–66, http://scx.sagepub.com/content/33/3/341.abstract. Allum found a +0.22 correlation between subjects' belief in astrology and their rating of the importance of "obedience" in children.

6. For example, the closest Mars ever gets to Earth is about 55 million kilometres, and Mars weighs about 6.4×10^{23} kilograms. So, a 50-kilogram doctor standing 0.5 metres from the mother exerts a gravitational force that is $(6.4 \times 10^{23} / 50) / (55 \times 10^9 / 0.5)^2 \doteq 0.99$ times as large—in other words, about the same.

7. See e.g. Robert Currey, "Empirical Astrology: Why It Is No Longer Acceptable to Say Astrology Is Rubbish on a Scientific Basis," http://www.astrology.co.uk/tests/basisofastrology.htm.

8. See e.g. Mary Regina Boland et al., "Birth Month Affects Lifetime Disease Risk: A Phenome-Wide Method," *Journal of the American Medical Informatics Association* 22, no. 1 (September 2015): 1042–53, http://jamia.oxfordjournals.org/content/early/2015/06/01/jamia.ocv046.

9. See e.g. Joshua K. Hartshorne, Nancy Salem-Hartshorne, and Timothy S. Hartshorne, "Birth Order Effects in the Formation of Long-Term Relationships," *Journal of Individual Psychology* 65, no. 2 (Summer 2009), and the related article Joshua K. Hartshorne, "How Birth Order Affects Your Personality," *Scientific American Mind*, January 1, 2010, http://www.scientificamerican.com/article/ruled-by-birth-order/.

10. Jacqueline Bigar, "Horoscope for Wednesday, July 27, 2016 [Libra]," *Toronto Star*, July 27, 2016, https://www.thestar.com/diversions/horoscope/2016/07/27/horoscope-for-wednesday-july-27-2016.html.

11. James Randi, *Flim-Flam!: Psychics, ESP, Unicorns, and Other Delusions* (Buffalo, NY: Prometheus, 1982), 61–62. The horoscopes were written for the Montreal publication *Midnight*, circa 1945, under the pen name "Zo-ran" (short for "Zodiacs by Randi").

12. Shawn Carlson, "A Double-Blind Test of Astrology," *Nature* 318, no. 6045 (December 5, 1985): 419–25. We focus here on "Part 2" of his experiment. See also "What Do You Mean, 'Test' Astrology?," *Skeptico*, February 16, 2005, http://skeptico.blogs.com/skeptico/2005/02/what_do_you_mea.html.

13. Specifically, all three profiles were ranked in terms of how well they corresponded to the subject, and the correct profile was ranked number one at a rate of 0.34 ± 0.044, number two at a rate of 0.40 ± 0.044, and number three at a rate of 0.25 ± 0.044.

14. J.D. McGervey, "A Statistical Test of Sun-Sign Astrology," *Zetetic* 1 (1977). Described in McGervey, *Probabilities in Everyday Life* (Chicago: Nelson-Hall, 1986), 45–46.

15. Indeed, this distribution passes the "chi-squared test" with p-value 0.0901, indicating that it is consistent with the star signs being completely random.

16. For example, Natalie Josef (in "The Best Career for Your Zodiac Sign," More, http://www.more.com/money/career-advice/best-career-your-zodiac-sign) and Excite Education (in "Best Careers according to Your Zodiac Sign," http://www.excite.com/education/blog/best-careers-according-to-your-zodiac-sign) both list "science" under Capricorn, Aquarius, and Scorpio only. A third article (Carol Stanley, "Astrology Signs: Best Careers for Each Zodiac Sign," Exemplore, https://exemplore.com/astrology/Astrology-Best-Professions-for-Each-Zodiac-Sign) does not mention "scientist" at all.

17. Emad Salib, "Astrological Birth Signs in Suicide: Hypothesis or Speculation?," *Medicine, Science, and the Law* 43, no. 2 (April 2003): 111–14, https://www.ncbi.nlm.nih.gov/pubmed/12741653.

18. This distribution passes the "chi-squared test" with $p=0.3063$.

19. Bernie I. Silverman and Marvin Whitmer, "Astrological Indicators of Personality," *Journal of Psychology* 87, no. 1 (1974): 89–95.

20. Alyssa Jayne Wyman and Stuart Vyse, "Science versus the Stars: A Double-Blind Test of the Validity of the NEO Five-Factor Inventory and Computer-Generated Astrological Natal Charts," *Journal of General Psychology* 135, no. 3 (July 2008): 287–300.

21. G.A. Tyson, "Occupation and Astrology or Season of Birth: A Myth?," *Journal of Social Psychology* 110, no. 1 (1980): 73–78.

22. For these students, the study found a chi-squared value of 21.93 with 11 degrees of freedom, corresponding to a p-value of 0.0249.

23. Dave Clarke, Toos Gabriels, and Joan Barnes, "Astrological Signs as Determinants of Extroversion and Emotionality: An Empirical Study," *Journal of Psychology* 130, no. 2 (1996): 131–40. For the one comparison of positive sun and moon signs versus

negative for both, they found a t-test value of 2.21 with 70 degrees of freedom, corresponding to a p-value of 0.015.

24. Peter Hartmann, Martin Reuter, and Helmuth Nyborg, "The Relationship between Date of Birth and Individual Differences in Personality and General Intelligence: A Large-Scale Study," *Personality and Individual Differences* 40, no. 7 (May 2006):1349–62.

25. Currey, "Empirical Astrology" (cited in note 7 above).

26. See e.g. Currey, "Empirical Astrology" and also Robert Currey, "U-Turn in Carlson's Astrology Test," *Correlation* 27, no. 2 (July 2011): 7–33, http://www.astrology -research.com/researchlibrary/.

27. See Chip Denman and Rick Adams, "JREF Status," James Randi Educational Foundation, September 1, 2015, http://web.randi.org/home/jref-status.

28. See e.g. James Randi, "Fakers and Innocents: The One Million Dollar Challenge and Those Who Try for It," *Skeptical Inquirer* 29, no. 4 (July/August 2005), http:// www.csicop.org/si/show/fakers_and_innocents_the_one_million_dollar_challenge _and_those_who_try_for.

29. See e.g. Adam Higginbotham, "The Unbelievable Skepticism of the Amazing Randi," *New York Times Magazine*, November 7, 2014, https://www.nytimes.com/ 2014/11/09/magazine/the-unbelievable-skepticism-of-the-amazing-randi.html.

30. See Robert Currey, "Astrology and James Randi," http://www.astrology.co.uk/tests/ randitest.htm.

31. A.J. Vicens, "Can the Zodiac Explain Why Washington, DC, Is So Messed Up?" *Mother Jones*, July/August 2014, http://www.motherjones.com/politics/2014/08/ zodiac-astrology-politicians-birthdays-elections.

32. My quick web search found the following assertions (and no others) about the best zodiac signs for politicians: "Careers suited to Aries: Politician" (Excite Education, "Best Careers"); "Aries . . . work well in the fields of government and politics" (Josef, "The Best Career"); "Politics: Aries, Gemini, Leo, Sagittarius" (Stanley, "Astrology Signs") and "Aries . . . Government and politics" ("What Career Should You Have According to Your Zodiac Sign?," Sun Gazing, http://www.sun-gazing.com/career -according-zodiac-sign/).

33. My quick web search found the following assertions (and no others) about the best zodiac signs for nurses: "Cancer . . . Nursing—Deal with drama and console those in pain" (Kim Evans, "The 4 Best Careers For Your Zodiac Sign," Jobs.net, http:// www.jobs.net/Article/CB-120-Talent-Network-Hospitality-The-4-Best-Careers- For-Your-Zodiac-Sign); "Pisces . . . Best jobs: . . .nurse" (Josef, "The Best Career"); "Career suited to Taurus: . . . Nurse" (Excite Education,"Best Careers"); "the intuitive qualities that are supposedly inherent to Pisces make us good for careers that involve compassion, like nursing . . ." (Lucia Peters, "What Job Should You Have Based on Your Zodiac Sign? This Infographic Might Tell You," Bustle, June 20, 2015, http://www.bustle.com/articles/90647-what-job-should-you-have-based-on -your-zodiacsign-this-infographic-might-tell-you); "[Aries] Many areas of medicine can be a great choice for you . . . nurse or surgeon . . . [Taurus] You'll feel happy working as nurse . . . [Gemini] . . . You can enjoy your busy shifts as a doctor or

a nurse.... [Pisces] ... You're a perfect candidate for jobs in healthcare. You can become a registered nurse, physician therapist or personal care aide. Any job that lets you connect with your patients is a good fit for you.... [Cancer] ... comfortable in jobs working directly with patients ... or a nursing assistant" ("What Your Zodiac Sign Says About Your Healthcare Career Choice," American Institute of Medical Sciences and Education, January 21, 2016, https://www.aimseducation. edu/blog/zodiac-sign-healthcare-career-choice/); "check out these sun signs which would best suit nursing: TAURUS, CANCER, LEO, VIRGO, LIBRA, SCORPIO, CAPRICORN, AQUARIUS, PISCES"; and Sun Gazing, "What Career" (cited above): "Pisces ... Nurse" (Find Your Fate, http://www.findyourfate.com/career/nursing.html).

34. I am grateful to Pauline Zvejnieks and Michael Hamilton-Jones for providing me with this data.

35. Data from Statistics Canada, giving the number of live births in Ontario on each date in the year 2012 (rounded to the nearest multiple of five).

36. If I make a 2×12 table of the two vectors of counts, and then use R's "chisq.test()" function to perform a chi-squared test of independence, it gives a p-value of 5.3×10^{-14}, which is extremely small and thus very highly significant.

37. See ShaoLan Hsueh's TED talk (cited in note 3 above).

38. See e.g. Mark Mayberry, "Astrology Fails the Test of Science," *Truth Magazine* 34, no. 18 (September 20, 1990): 560–63, http://www.truthmagazine.com/archives/volume34/GOT034263.html.

39. See e.g. chapter 10 of H.J. Eysenck and D.K.B. Nias, *Astrology: Science or Superstition?* (London: Temple Smith, 1982).

40. See e.g. Claude Benski et al., *The "Mars Effect": A French Test of Over 1,000 Sports Champions* (Amherst, NY: Prometheus, 1996), or the summary available at https://www.amazon.com/Mars-Effect-Claude-Benski/dp/0879759887.

41. Paul Kurtz, Marvin Zelem, and George Abell, "Results of the U.S. Test of the 'Mars Effect' Are Negative," *Skeptical Inquirer* 4, no. 2 (Winter 1979–1980): 19–26.

42. See Dennis Rawlins, "sTARBABY," *Fate*, October 1981, 67–98, http://cura.free.fr/xv/14starbb.html. See also the following responses to these allegations: J.J. Lippard, "Mars Effect (Re: 'Crybaby')," sci.skeptic newsgroup, January 20, 1992, available at https://www.discord.org/~lippard/jjl-on-crybaby.txt; and J.J. Lippard, "Skeptics and the 'Mars Effect': A Chronology of Events and Publications," June 25, 2016, available at https://www.discord.org/~lippard/mars-effect-chron.rtf. See also Kenneth Irving, "A Brief Chronology of the 'Mars Effect' Controversy", Planetos.info, http://www.planetos.info/marchron.html.

43. See e.g. the quotation on Quotes.net: http://www.quotes.net/mquote/818023. To view a scene from the series, see "Numb3rs Scene: Everything Is Numbers, Math Is Everywhere," YouTube, uploaded November 15, 2007, https://www.youtube.com/watch?v=vFRTgr7MfWw.

44. The numerology reading generated by the magazine *So Feminine* for my own birth date of October 13, 1967, can be found at http://www.sofeminine.co.uk/astro/

numerologie/07metiers/07metiers1.asp?j=13&m=10&a=1967&Submit=Enter.

45. See Corrine Lane, "Zodiac Sign Found Most among U.S. Presidents," *Astrology Blog*, Astrology Library, March 18, 2016, https://astrolibrary.org/zodiac-sign-us -presidents/.

Chapter 21: Mind over Matter?

1. For discussion and video, see e.g. Kirk Zamieroski, "How Do Optical Illusions Work?," Inside Science, July 29, 2015, https://www.insidescience.org/video/ how-do-optical-illusions-work.

2. See "James Randi Debunks Peter Popoff Faith Healer," YouTube, uploaded May 19, 2006, https://www.youtube.com/watch?v=q7BQKu0YP8Y.

3. See Robert Todd Carroll, "Project Alpha," *Skeptic's Dictionary*, http://www.skepdic .com/projectalpha.html.

4. See e.g. Norman D. Sundberg, "The Acceptability of 'Fake' versus 'Bona Fide' Personality Test Interpretations," *Journal of Abnormal and Social Psychology* 50, no. 1 (February 1955): 145–57; C.R. Snyder and R.J. Shenkel, "The P.T. Barnum Effect," *Psychology Today* 8, no. 10 (1975): 52–54; and Ray Hyman, "Cold Reading: How to Convince Strangers that You Know All about Them," in *Paranormal Borderlands of Science*, ed. Kendrick Frazier (Buffalo, NY: Prometheus, 1981), 79–96.

5. Bertram R. Forer, "The Fallacy of Personal Validation: A Classroom Demonstration of Gullibility," *Journal of Abnormal and Social Psychology* 44, no. 1 (January 1949): 118–23, http://apsychoserver.psych.arizona.edu/JJBAReprints/PSYC621/Forer_The fallacy of personal validation_1949.pdf.

6. Available at http://starecat.com/you-will-continue-to-interpret-vague-statements -as-uniquely-meaningful-chinese-fortune-cookie-quote/.

7. Daryl J. Bem, "Feeling the Future: Experimental Evidence for Anomalous Retroactive Influences on Cognition and Affect," *Journal of Personality and Social Psychology* 100, no. 3 (March 2011): 407–25. A version is available online at http:// dbem.org/FeelingFuture.pdf.

8. In "Experiment 1" of his paper.

9. Charles M. Judd and Bertram Gawronski, "Editorial Comment," *Journal of Personality and Social Psychology* 100, no. 3 (March 2011): 406, http://psycnet.apa .org/journals/psp/100/3/406/.

10. See e.g. "Newton's Laws of Motion," Glenn Research Center, NASA, https://www.grc .nasa.gov/www/k-12/airplane/newton.html, or any introductory physics textbook.

11. "Determinism is the philosophical belief that every event or action is the inevi-table result of preceding events and actions. Thus, in principle at least, every event or action can be completely predicted in advance, or in retrospect." Matthew A. Trump, "Lesson One: The Philosophy of Determinism," http://order.ph.utexas.edu/ chaos/determinism.html.

12. See e.g. Elizabeth Howell, "Time Travel: Theories, Paradoxes & Possibilities," Space.com, June 21, 2013, https://www.space.com/21675-time-travel.html.

13. See e.g. "The Wave Function as a Probability," at http://physicspages.com/pdf/ Griffiths%20QM/Wave%20function%20as%20probability.pdf, or any introductory quantum mechanics textbook.

14. See e.g. P.C.W. Davies, "Quantum Tunneling Time," *American Journal of Physics* 73, no. 1 (January 2005): 23–27, or any introductory quantum mechanics textbook.

15. See e.g. "Sorry, Einstein—Physicists Just Reinforced the Reality of Quantum Weirdness in the Universe," *Science Alert*, February 8, 2017, https://www .sciencealert.com/sorry-einstein-physicists-just-reinforced-the-reality-of -quantum-weirdness-in-the-universe, or Amir D. Aczel, *Entanglement* (New York: Four Walls Eight Windows, 2002).

16. See e.g. Lisa Zyga, "Physicists Provide Support for Retrocausal Quantum Theory, in Which the Future Influences the Past," Phys.org, July 5, 2017, https://phys .org/news/2017-07-physicists-retrocausal-quantum-theory-future.html; Matthew S. Leifer and Matthew F. Pusey, "Is a Time Symmetric Interpretation of Quantum Theory Possible without Retrocausality?," *Proceedings of the Royal Society A* 473, no. 2202 (June 2017), http://rspa.royalsocietypublishing.org/content/473/2202/ 20160607; Mike McRae, "This Quantum Theory Predicts that the Future Might Be Influencing the Past," *Science Alert*, July 6, 2017, https://www.sciencealert.com/this -quantum-theory-predicts-the-future-might-influence-the-past; and the extensive discussion in David Ellerman, "A Very Common Fallacy in Quantum Mechanics: Superposition, Delayed Choice, Quantum Erasers, Retrocausality, and All That," preprint submitted December 16, 2011, https://arxiv.org/abs/1112.4522.

17. See e.g. Jennifer Ouellette, "Can Quantum Physics Explain Consciousness?," *Atlantic*, November 7, 2016, https://www.theatlantic.com/science/archive/2016/11/ quantum-brain/506768/.

18. The researcher is Gergö Hadlaczky; see his page on the Karolinska Institutet website at http://ki.se/en/people/gerhad. His experiments are reported in his 2003 paper "Precognitive Habituation: An Attempt to Replicate Previous Results," available at https://www.researchgate.net/publication/223467682_Precognitive_ habituation_An_attempt_to_replicate_previous_results.

19. Jeff Galak, Robyn A. LeBoeuf, Leif D. Nelson, and Joseph P. Simmons, "Correcting the Past: Failures to Replicate Psi," *Journal of Personality and Social Psychology* 103, no. 6 (December 2012): 933–48, https://papers.ssrn.com/sol3/papers.cfm?abstract_ id=2001721.

20. Stuart J. Ritchie, Richard Wiseman, and Christopher C. French, "Failing the Future: Three Unsuccessful Attempts to Replicate Bem's 'Retroactive Facilitation of Recall' Effect," *PLoS One* 7, no. 3 (March 2012): e33423, http://journals.plos.org/plosone/ article?id=10.1371/journal.pone.0033423.

21. See Peter Aldhous, "Journal Rejects Studies Contradicting Precognition," *New Scientist Daily News*, May 5, 2011, https://www.newscientist.com/article/ dn20447-journal-rejects-studies-contradicting-precognition. See also Stuart J. Ritchie, Richard Wiseman, and Christopher C. French, "Replication, Replication,

Replication," *Psychologist* 25, no. 5 (May 2012): 346–57, https://thepsychologist.bps .org.uk/volume-25/edition-5/replication-replication-replication.

22. See e.g. Lea Winerman, "Interesting Results: Can They Be Replicated?," *Monitor on Psychology* 44, no. 2 (February 2013): 38, http://www.apa.org/monitor/2013/02/ results.aspx.

23. Daryl J. Bem, Patrizio Tressoldi, Thomas Rabeyron, and Michael Duggan, "Feeling the Future: A Meta-analysis of 90 Experiments on the Anomalous Anticipation of Random Future Events" (unpublished paper, 2014), http://dbem.org/FF%20Meta -analysis%206.2.pdf.

24. See e.g. E.J. Wagenmakers's review "Bem Is Back: A Skeptic's Review of a Meta-analysis on Psi," Open Science Collaboration, June 25, 2014, http://osc .centerforopenscience.org/2014/06/25/a-skeptics-review/.

25. See the Center for the Study of Emotion and Attention at the University of Florida (http://csea.phhp.ufl.edu/).

26. As specified in Margaret M. Bradley and Peter J. Lang, "IAPS Message," Center for the Study of Emotion and Attention, University of Florida, http://csea.phhp.ufl .edu/media/iapsmessage.html.

27. See Associated Press, "Twins Give Birth Minutes Apart in Same Hospital," *Today*, NBC News, December 22, 2011, https://www.today.com/news/twins-give-birth -minutes-apart-same-hospital-wbna45769823.

28. Michael Betcherman, *Face-off* (Toronto: Razorbill, 2014). A description can be found on the publisher's website at https://penguinrandomhouse.ca/ books/392207/.

29. The joke appeared (though it was about brothers, not twins) in Woody Allen, *Without Feathers* (New York: Ballantine, 1986).

30. See e.g. Karen Kirkpatrick, "Can Twins Sense Each Other?," *How Stuff Works*, July 17, 2015, https://science.howstuffworks.com/life/genetic/can-twins-sense -each-other.htm and the related posting "Is 'Twin Communication' a Real Thing?," *The Body Odd*, NBC News, December 28, 2011, http://bodyodd.nbcnews.com/_ news/2011/12/28/9750598-is-twin-communication-a-real-thing.

31. Susan J. Blackmore and Frances Chamberlain, "ESP and Thought Concordance in Twins: A Method of Comparison," *Journal of the Society for Psychical Research* 59, no. 831 (1993): 89–96.

32. See "30 Priceless Quotes Said by Robin Williams. Truly a Legend," *Tickld*, January 19, 2018, http://www.tickld.com/x/fbk/30-priceless-quotes-said-by-robin-williams -truly-a-legend/p-26.

33. See e.g. "Steven Wright Quotes," BrainyQuote.com, https://www.brainyquote.com/ quotes/quotes/s/stevenwrig578926.html.

34. See e.g. "Fact or Fiction?," NewEarthArmy.com, http://neweartharmy.com/Fact_ or_Fiction.html.

35. Related US Army–funded programs on remote viewing prior to Stargate apparently included code names such as SCANATE, SRI, ACSI, SED, Gondola Wish,

Grill Flame, INSCOM, ICLP, Sun Streak, and SAIC. See e.g. "Star Gate (Controlled Remote Viewing)," *Intelligence Resource Program*, Federation of American Scientists, https://fas.org/irp/program/collect/stargate.htm.

36. The report (Michael D. Mumford, Andrew M. Rose, and David A. Goslin, *An Evaluation of Remote Viewing: Research and Applications*, prepared by the American Institutes for Research, September 29, 1995) is available at http://www.lfr.org/wp-content/uploads/2017/02/AirReport.pdf or https://www.cia.gov/library/readingroom/document/cia-rdp96-00791r000200180006-4.

37. See Committee of Presidents of Statistical Societies, news release, August 1, 2007, archived at http://probability.ca/jeff/copssaward. A photo of Utts presenting me with the award can be seen at http://probability.ca/jeff/images/copssaward.jpg.

38. See Allan Rossman, "Interview with Jessica Utts," *Journal of Statistics Education* 22, no. 2 (2014), http://ww2.amstat.org/publications/jse/v22n2/rossmanint.pdf. At the top of page 20, she mentions being born on Saturday.

39. See Jessica Utts, "Appreciating Statistics," *Journal of the American Statistical Association* 111, no. 516 (2016): 1373–80. The quoted statement appears on page 1379.

40. See e.g. Richard Wiseman and Julie Milton, "Experiment One of the SAIC Remote Viewing Program: A Critical Re-evaluation," *Journal of Parapsychology* 62, no. 4 (December 1998): 297–308, available at http://www.richardwiseman.com/resources/SAICcrit.pdf.

41. See the program's web page archived at https://web.archive.org/web/20180329071828/www.princeton.edu/~pear.

42. See "Experimental Research: I. Human-Machine Anomalies," Princeton Engineering Anomalies Research," archived at https://web.archive.org/web/20171206203522/http://www.princeton.edu/~pear/experiments.html.

43. John McCrone, quoted in Robert Todd Carroll, "The Princeton Engineering Anomalies Research (PEAR)," *Skeptic's Dictionary*, http://skepdic.com/pear.html. McCrone's paper, "Psychic Powers: What Are the Odds?," appeared in *New Scientist* in November 1994.

44. See Stanley Jeffers, "The PEAR Proposition: Fact or Fallacy?," *Skeptical Inquirer* 30, no. 3 (May/June 2006), http://www.csicop.org/si/show/pear_proposition_fact_or_fallacy.

45. See C.E.M. Hansel, *The Search for Psychic Power: ESP and Parapsychology Revisited* (Buffalo, NY: Prometheus, 1989), which is quoted in the *Skeptic's Dictionary* article on PEAR.

46. See R. Jahn et al., "Mind/Machine Interaction Consortium: PortREG Replication Experiments," *Journal of Scientific Exploration* 14, no. 4 (2000): 499–555, archived at https://web.archive.org/web/20171130193844/https://www.princeton.edu/~pear/pdfs/2000-mmi-consortium-portreg-replication.pdf.

47. See e.g. George P. Hansen, Jessica Utts, and Betty Markwick, "Critique of the PEAR Remote-Viewing Experiments," *Journal of Parapsychology* 56, no. 2 (June 1992): 97–113, http://www.tricksterbook.com/ArticlesOnline/PEARCritique.htm.

48. See Associated Press, "Report: Princeton to Close ESP Lab," *USA Today*, February 11, 2007, http://usatoday30.usatoday.com/news/education/2007-02-11-princeton -esp_x.htm.

49. "Remote Viewing," Ministry of Defence, available through the UK National Archives at http://webarchive.nationalarchives.gov.uk/20121026065214/ http://www.mod.uk/DefenceInternet/FreedomOfInformation/DisclosureLog/ SearchDisclosureLog/RemoteViewing.htm. For media reports, see e.g. "MoD Defends Psychic Powers Study," BBC News, February 23, 2007, http://news.bbc .co.uk/2/hi/uk_news/6388575.stm and "Defence Chiefs Spent £18,000 on a Mystic Experiment to Find bin Laden's Lair," *Evening Standard* (London), February 24, 2007, http://www.standard.co.uk/news/defence-chiefs-spent-18000-on-a-mystic -experiment-to-find-bin-ladens-lair-7085768.html.

Chapter 22: Lord of the Luck

1. See e.g. "The Global Religious Landscape," Pew Research Center, December 18, 2012, http://www.pewforum.org/2012/12/18/global-religious-landscape-exec, which estimated that in 2010, 84 percent of the world's population had some reli- gious affiliation.

2. See e.g. the cached image at https://s-media-cache-ak0.pinimg.com/736x/f8/30/75/ f83075f25f6845ba9a4de1eb3687b1c8--snoopy-charlie-snoopy-peanuts.jpg.

3. See e.g. Bodie Hodge, "Chapter 4: How Old Is the Earth?," *New Answers Book 2*, Answers in Genesis, https://answersingenesis.org/age-of-the-earth/how-old-is-the-earth/.

4. Quoted by a viewer commenting on the Rotten Tomatoes website. See "Thirteen Conversations about One Thing Reviews," Rotten Tomatoes, January 9, 2011, https://www.rottentomatoes.com/m/thirteen_conversations_about_one_thing/ reviews/?page=2&type=user.

5. The character Lloyd, played by Jack Warden, in a scene written and directed by Woody Allen. See "September (1987) Woody Allen: '. . . Haphazard, Morally Neutral and Unimaginably Violent . . . ," YouTube, uploaded April 4, 2010, https:// www.youtube.com/watch?v=kW-drCJhqSE.

6. See "Banana: The Athiests [*sic*] Nightmare," YouTube, uploaded June 4, 2006, https://www.youtube.com/watch?v=nfv-Qn1M58I.

7. See e.g. Vanessa Richins Myers, "Do Bananas Have Seeds?," *Spruce*, https://www .thespruce.com/do-bananas-have-seeds-3269378.

8. See RTÉ—Ireland's National Public Service Media, "Stephen Fry on God—The Meaning of Life," YouTube, uploaded January 28, 2015, https://www.youtube.com/ watch?v=-suvkwNYSQo.

9. See e.g. Leonard Greene, "Steve Harvey Announces Wrong Miss Universe Winner," *New York Daily News*, December 21, 2015, http://www.nydailynews.com/ entertainment/steve-harvey-announces-wrong-universe-winner-article-1.2472285. For video of the incident, see e.g. "Steve Harvey Announces the Wrong Winner of Miss Universe 2015," YouTube, uploaded December 20, 2015, https://www.youtube .com/watch?v=3DKDaSd-4nY.

10. See "Steve Harvey on Atheism!," YouTube, uploaded January 22, 2015, https://www.youtube.com/watch?v=VWJ9ylZkS2s.

11. "According to . . . the recently published *Oxford Handbook of Atheism*, there are approximately 450–500 million non-believers in God worldwide." Phil Zuckerman, "How Many Atheists Are There?," *Psychology Today*, October 20, 2015, https://www.psychologytoday.com/blog/the-secular-life/201510/how-many-atheists-are-there. See also Ariela Keysar and Juhem Navarro-Rivera, "A World of Atheism: Global Demographics," in *The Oxford Handbook of Atheism*, ed. Stephen Bullivant and Michael Ruse (Oxford: Oxford University Press, 2013), 553–86.

12. See Edward J. Larson and Larry Witham, "Leading Scientists Still Reject God," letter to the editor, *Nature* 394 (July 23, 1998): 313, http://www.nature.com/nature/journal/v394/n6691/full/394313a0.html.

13. See Michael Stirrat and R. Elisabeth Cornwell, "Eminent Scientists Reject the Supernatural: A Survey of the Fellows of the Royal Society," *Evolution: Education and Outreach* 6, no. 1 (December 2013): 33, https://link.springer.com/article/10.1186/1936-6434-6-33.

14. See e.g. "27 Celebrities You Probably Didn't Know Are Atheists," Think Atheist, July 2, 2009, http://www.thinkatheist.com/profiles/blogs/27-celebrities-you-probably.

15. See e.g. David Willey, "Vatican 'Must Immediately Remove' Child Abusers—UN," BBC News, February 5, 2014, http://www.bbc.com/news/world-europe-26044852.

16. See e.g. Stephanie Kirchgaessner and Melissa Davey, "George Pell Takes Leave from Vatican to Fight Sexual Abuse Charges in Australia," *Guardian* (London), June 29, 2017, https://www.theguardian.com/australia-news/2017/jun/29/george-pell-takes-leave-from-vatican-to-fight-sex-abuse-charges-in-australia.

17. See e.g. "Secondary Wars and Atrocities of the Twentieth Century: India (1947)," Necrometrics.com, http://necrometrics.com/20c300k.htm#India.

18. See e.g. Malcolm Sutton, "An Index of Deaths from the Conflict in Ireland," CAIN (Conflict Archive on the Internet), Ulster University, http://cain.ulst.ac.uk/sutton/tables/Status.html.

19. See Julia Marsh, "Some Victims' Funerals Will Be Held at Gunman's Church," *New York Post*, December 17, 2012, http://nypost.com/2012/12/17/some-victims-funerals-will-be-held-at-gunmans-church/.

20. In a sample of 1,208 university psychology students, 7 percent answered yes to the statement "If God told me to kill I would do it in His name"; see M.A. Persinger, "'I Would Kill in God's Name': Role of Sex, Weekly Church Attendance, Report of a Religious Experience, and Limbic Lability," *Perceptual and Motor Skills* 85, no. 1 (1997): 128–30 and Michael A. Persinger, "Variables that Predict Affirmative Responses to the Item If God Told Me to Kill I Would Do It in His Name: Implications for Radical Religious Behaviours," *Journal of Socialomics* 5, no. 3 (2016): e166, https://www.omicsgroup.org/journals/variables-that-predict-affirmative-responses-to-the-item-if-god-told-meto-kill-i-would-do-it-in-his-name-implications-for-radical-2471-8726-1000166.php?aid=73879.

21. See also e.g. "10 People Who Give Christianity a Bad Name," Listverse, February 23, 2010, http://listverse.com/2010/02/23/10-people-who-give-christianity-a-bad-name/.

22. In an interview, when asked, "Do you believe in God?" Pitt replied, "No, no, no!" Norbert Körzdörfer, "'With Six Kids Each Morning It Is about Surviving!,'" *Bild* (Berlin), July 29, 2009, http://www.bild.de/news/bild-english/inglourious-basterd -star-on-angelina-jolie-and-six-kids-9110388.bild.html. See also Gina Salamone, "Brad Pitt: 'I'm Probably 20 Percent Atheist and 80 Percent Agnostic,'" *New York Daily News*, July 23, 2009, http://www.nydailynews.com/entertainment/gossip/ brad-pitt-20-percent-atheist-80-percent-agnostic-article-1.394661; "Brad Pitt," *Celebrity Atheist List*, http://www.celebatheists.com/wiki/Brad_Pitt; and "Angelina Jolie," *Celebrity Atheist List*, http://www.celebatheists.com/wiki/Angelina_Jolie.

23. See e.g. "Bono, Brad Pitt Launch Campaign for Third World Relief," MTV News, April 6, 2005, http://www.mtv.com/news/1499708/bono-brad-pitt-launch -campaign-for-third-world-relief/; "Brad & Angelina Start Charitable Group," *People*, September 20, 2006, http://people.com/celebrity/brad-angelina-start -charitable-group/; and Roger Friedman, "Angelina Jolie and Brad Pitt's Charity: Bravo," Fox News, March 21, 2006, http://www.foxnews.com/story/2008/03/21/ angelina-jolie-and-brad-pitt-charity-bravo.html?sPage=fnc/entertainment/ celebrity/pitt.

24. See e.g. the interesting video excerpt: BerkshireInsurance, "Warren Buffett on Spiritualism God and Rebirth," YouTube, uploaded June 27, 2012, https://www .youtube.com/watch?v=ZNWX0CZm3lk.

25. See e.g. "Warren Buffett," *Wall Street Donors Guide*, Inside Philanthropy, https:// www.insidephilanthropy.com/wall-street-donors/warren-buffett.html and Chase Peterson-Withorn, "Warren Buffett Just Donated Nearly $2.9 Billion to Charity," *Forbes*, July 14, 2016, https://www.forbes.com/sites/chasewithorn/2016/07/14/ warren-buffett-just-donated-nearly-2-9-billion-to-charity/.

26. For analysis, discussion, and related links, see Hemant Mehta, "Are Religious People Really More Generous than Atheists? A New Study Puts That Myth to Rest," *Friendly Atheist*, November 28, 2013, http://www.patheos.com/blogs/ friendlyatheist/2013/11/28/are-religious-people-really-more-generous-than -atheists-a-new-study-puts-that-myth-to-rest/ and Jay Michaelson, "New Study: Three-quarters of American Giving Goes to Religion," *Religion Dispatches*, December 12, 2013, http://religiondispatches.org/new-study-three-quarters-of -american-giving-goes-to-religion/.

27. Jean Decety, "The Negative Association between Religiousness and Children's Altruism across the World," *Current Biology* 25, no. 22 (November 16, 2015): 2951–55, http://www.cell.com/current-biology/abstract/S0960-9822(15)01167 -7. For media coverage of Decety's paper, see e.g. Helena Horton, "Muslims and Christians Less Generous than Atheists, Study Finds," *Telegraph* (London), November 6, 2015, http://www.telegraph.co.uk/news/religion/11979235/ Muslims-and-Christians-less-generous-than-atheists-study-finds.html and

Warren Cornwall, "Nonreligious Children Are More Generous," *Science*, November 5, 2015, http://www.sciencemag.org/news/2015/11/nonreligious -children-are-more-generous.

28. See the Family Research Council's website: http://www.frc.org/about-frc.
29. From "Homosexuality," Family Research Council, http://www.frc.org/homosexuality.
30. The interview, with conservative rabbi Jonathan Cahn, can be heard at Brian Tashman, "Jonathan Cahn: Hurricane Joaquin May Hit DC as Punishment for Gay Marriage," Right Wing Watch, October 5, 2015, http://www.rightwingwatch.org/ post/jonathan-cahn-hurricane-joaquin-may-hit-dc-as-punishment-for-gay -marriage/. It can also be accessed on Soundcloud: https://soundcloud.com/ rightwingwatch/cahn-hurricane-joaquin-may-hit-dc-as-punishment-for-gay -marriage. Also, see e.g. AllenMcw, "FRC Tony Perkins & Jonathan Cahn Claimed Joaquin Will Hit DC as Punishment for Marriage Equality," *Daily Kos*, October 5, 2015, https://www.dailykos.com/stories/2015/10/5/1428159/-FRC-Tony -Perkins-Jonathan-Cahn-claimed-Joaquin-will-hit-DC-as-Punishment-for -Marriage-Equality and John Paul Brammer, "Tony Perkins Blamed Gay People for God's Wrath. His House Was Swept Away," *Guardian* (London), August 18, 2016, https://www.theguardian.com/commentisfree/2016/aug/18/tony -perkins-floods-louisiana-gay-christian-conservative.
31. See e.g. Kate Nelson, "Louisiana Floods Destroy Home of Christian Leader Who Says God Sends Natural Disasters to Punish Gay People," *Independent* (London), August 18, 2016, http://www.independent.co.uk/news/world/americas/christian -home-destroyed-flood-tony-perkins-natural-disasters-gods-punishment -homosexuality-a7196786.html; Michael Baggs, "US Pastor, Who Believes Floods Are God's Punishment, Flees Flooded Home," *Newsbeat*, BBC News, August 18, 2016, http://www.bbc.co.uk/newsbeat/article/37116661/us-pastor-who-believes -floods-are-gods-punishment-flees-flooded-home; Sky Palma, "Guy Who Says God Sends Natural Disasters to Punish Gays Has His Home Destroyed in a Natural Disaster," *DeadState*, August 17, 2016, http://deadstate.org/guy-who-says-god -sends-natural-disasters-to-punish-gays-has-his-home-destroyed-in-a-natural -disaster/; "'God Is Trying to Send Us a Message': Pastor Who Believes God Wants to Punish Gays Driven from Home by Floods," *National Post* (Toronto), August 19, 2016, http://nationalpost.com/news/world/god-is-trying-to-send-us-a-message -pastor-who-believes-god-wants-to-punish-gays-driven-from-home-by-floods; Jack Holmes, "A Man Who Says God Punishes Gays with Natural Disasters Had His Home Destroyed in the Flood," *Esquire*, August 18, 2016, http://www.esquire. com/news-politics/news/a47783/tony-perkins-anti-gay-flood/; and the August 16, 2016, episode of Perkins's own podcast, *Washington Watch* (https://soundcloud .com/family-research-council/20160816-tony-perkins).
32. Albert Einstein, "Religion and Science," *New York Times Magazine*, November 9, 1930, 1–4, available at http://www.sacred-texts.com/aor/einstein/einsci.htm.
33. For starters, see e.g. "Lines of Evidence: The Science of Evolution," *Understanding Evolution*, http://evolution.berkeley.edu/evolibrary/article/lines_01 and Stated

Clearly, "What Is the Evidence for Evolution?," YouTube, uploaded October 10, 2014, https://www.youtube.com/watch?v=lIEoO5KdPvg.

34. See e.g. "Carl Sagan '100 Billion Galaxies Each with 100 Billion Stars,'" YouTube, uploaded February 26, 2008, https://www.youtube.com/watch?v=5Ex__M-OwSA. In fact, Sagan talked about this so often that Johnny Carson satirized him using the phrase "billions and billions," though apparently Sagan never actually said that phrase—see "Carl Sagan Takes Questions: More from His 'Wonder and Skepticism' CSICOP 1994 Keynote," *Skeptical Inquirer* 29, no. 4 (July/August 2005), http://www.csicop.org/si/show/carl_sagan_takes_questions.

35. See e.g. Fraser Cain, "How Many Stars Are There in the Universe?," *Universe Today*, June 3, 2013, https://www.universetoday.com/102630/how-many-stars-are-there-in-the-universe/.

36. See e.g. the ongoing catalogue of exoplanets at http://exoplanet.eu/catalog/.

Chapter 23: Lucky Reflections

1. See e.g. "Heaven Can Wait (7/8) Movie Clip—How Heaven Works (1978) HD," YouTube, uploaded October 11, 2011, https://www.youtube.com/watch?v=SzVAyGry2Ic.

2. See the movie's entry on IMDB at http://www.imdb.com/title/tt3758708/.

3. Stephen Fry, *The Hippopotamus: A Novel* (New York: Soho Press, 2014). See http://www.penguinrandomhouse.com/books/56854/hippopotamus-by-stephen-fry/.

4. See e.g. "Pancreatic Cancer Survival Rates, by Stage," American Cancer Society, https://www.cancer.org/cancer/pancreatic-cancer/detection-diagnosis-staging/survival-rates.html.

5. See her obituary at "Helen Stephanie Rosenthal," Legacy.com, http://www.legacy.com/obituaries/thestar/obituary.aspx?n=helen-stephanie-rosenthal&pid=186480364.

Index

acupuncture, 68
Adams, Douglas, 241
airplane crashes, 111, 192
Alabama, 189–190
albatrosses, 13
Alberta, 58, 204
Aleichem, Sholem, 25
allergies, 121–122
alternate cause (luck trap), 51, 278
alternative medicine, 128–130
Alvergne, Alexandra, 81
anecdotes, 128–130
Anthony, Stephone, 182
aphorisms, 192–202
Arkin, Alan, 72, 75, 253
asthma, 121–122
astrology, 1, 5, 214–234
atheism, 255–263
Atlanta (hockey team), 179

atomic bombs, 17–19, 112–113
autism, 137

B-52 (fighter plane), 112–113
backgammon, 40, 127–128
Bacon, Kevin, 259
Balder the Beautiful, 25
Baltimore, 203
bananas, 168–169, 255–256
baseball, 36, 40–42, 69, 177–179,
    180–182, 195
Baseball Hall of Fame, 180
basketball, 131–132, 173–177, 182
Battle of Stalingrad, 22
Bayesian inference, 119, 144
bears, 112
Beat the Streak, 181
Beatty, Warren, 35, 264
Beckinsale, Kate, 42

beginner's luck, 182–183

Bem, Daryl, 239–243

biased observation (luck trap), 52, 278

big bang, 251

birthday wishes, 44–45

Björk, 259

Blue Jays (baseball team). *See* Toronto
  Blue Jays

Bogart, Humphrey, 35

bombs, atomic, 17–19, 112–113

Bond, James, 132

Boone, Aaron, 41

Booth, John Wilkes, 87

Boston Red Sox (baseball team), 40–41

bracketology, 174

breast cancer, 75, 129–130

Brexit, 184–185

British Medical Journal, 29–30, 137

British Museum, 17

Bublé, Michael, 21

Buffett, Warren, 23, 159, 174, 260

Burns, Robbie, 6

Byrne, Rhonda, 73

Calgary, 84–86

cancer, 21, 75, 114, 129–130, 266,
  268–269

Carlson, Shawn, 218

Carpenter, Mary Chapin, 148

Carrey, Jim, 72

Casablanca (movie), 35

Cashore, Harvey, 152

casinos, 39, 60, 91, 133–135

Castro, Fidel, 26

Catcher in the Rye, The (novel), 22

Caulfield, Holden, 22

CBC, 152

celiac disease, 138

CIA, 246

cereal, 67–69

Channon, Jim, 246

Chase, Chevy, 105

Chekhov, Anton, 47

chemotherapy, 129, 268

Chicago Cubs (baseball team), 42

China, 26, 214, 230

Christie, Julie, 35

clairvoyance, 245–249

Clark, Sally, 208–210

Clarke, Bernice, 82–84

Cleaver, Thomas McKelvey, 196

climate change, 124–126

Clinton, Hillary, 45, 188–189

Clooney, George, 246

clovers, four-leaf. *See* four-leaf clovers

Cobb, Ty, 181

Coboconk, Ontario, 152

coffee, 142

coincidences, 11, 14–16, 55–56, 62–66,
  69–71, 82–90

Coleridge, Samuel Taylor, 13

Collins, Janet, 207

Color Purple, The (movie), 73

comedy, improvisational, 165–167

Comfort, Ray, 255

Corbett, Boston, 89

Costner, Kevin, 36

cot death. *See* SIDS

CPR (cardiopulmonary resuscitation),
  69–71

craps (game), 8–9, 58–60, 132–133

crib death. *See* SIDS

crime, 105–110, 113–114, 203–213

Crimean War, 116

crime rates, 107–109, 113–114

crossing fingers, 12

Cumberbatch, Benedict, 35

cup of coffee (expression), 182

Curry, Stephen, 131
Curse of the Bambino, 40–41, 173
Cusack, John, 42

Dahlen, Bill, 181
Dallas, 87, 88
Damon, Matt, 72
Deal or No Deal (TV show), 164–165
de Berk, Lucia, 211–213
dermatitis, 122
determinism, 240–241
different meaning (luck trap), 51, 278
DiMaggio, Joe, 180–182
Dirty Harry (movie), 21
dreams, 10, 60–62
Dr. Strange (movie), 35
Duke (basketball team), 175
dumb luck, 2, 277
Dunkirk, Martin, 125
Dupilumab, 122
Dylan, Bob, 194
dyscrasia, 139
Dyson, Chris, 84

Eastwood, Clint, 21
Edmonds, Bob, 152
Edmonton Institution for Women, 204
Einstein, Albert, 240, 261
election polls, 45, 184–191
electoral college, 187–188
Elizabeth II, Queen, 26
Epps, Charlie, 232
escape room (game), 204
ESP (extra-sensory perception), 9–10,
    60–62, 244–249
EuroMillions (lottery), 158
European Union, 184
European Vacation (movie), 105
evolution, 68, 79, 251, 256, 257, 262–263

extended warranties, 135, 136
extra-large target (luck trap), 50, 278
extra-sensory perception (ESP). *See* ESP

Facebook, 85, 86
facts that go together (luck trap), 53, 278
faith healing, 236–237
false reporting (luck trap), 51, 278
Family Research Council, 260
faster-than-light travel, 240
Fat Man (bomb), 17–19
Feist (musician), 26
fertility, 12, 77–78
fiction, 35–38, 46–47, 200–201, 235, 241,
    264–268
Fiddler on the Roof (musical), 25
Field of Dreams (movie), 36
Fifth Estate, The (TV show), 152
fingers crossed, 12
Finland, 30
First Earth Battalion, 246
Fisher, Danielle and Nicole, 244
Flaig, Steve, 16
forceful luck, 3, 277
Ford (motor company), 87
Ford's Theatre, 87
Forer, Bertram, 238
fortune cookies, 238
Foster, Jodie, 259
four-leaf clovers, 3, 12
Foxx, Jamie, 72
Friday the Thirteenth, 24–31
friggatriskaidekaphobia, 25
Fry, Stephen, 257–258, 268
Fuller, Buckminster, 201

Gallup, George, 187–188
gambling, 9, 58–60, 132–136
Gauquelin, Michel, 230–232

gay marriage, 261
general theory of relativity, 240–242
Georgia, 9, 59
Gere, Richard, 72
germs, 138–140, 216
Gillett, Lamar, 195
global warming, 124–126
gluten, 138
Golden Globe Awards, 74
Goldfinger, Auric, 132
Goldsboro, North Carolina, 112
Gomez, Lefty, 195
Good, the Bad, and the Ugly, The
    (movie), 21
Gordon, Avrum, 84
Gore, Al, 125
Gorski, David H., 129
Grambling State University (basketball
    team), 176
Greenwich Village, 39
Guardian (newspaper), 185
Guerriero, Linda, 152
Guinness World Records, 71

Haghani, Victor, 124
Halligan, Dickie, 15
Harvard (basketball team), 177
Harvey, Steve, 72, 258–259, 262
haunted houses, 39, 92, 104
Hawaii, 14–15, 55, 135
Heaven Can Wait (movie), 35, 264
Heimlich manoeuver, 69
Heinz Company, 182
hidden help (luck trap), 50, 278
Hill, Rick, 15
Hippopotamus, The (movie), 266
Hiroshima, 17, 112
Hitchhiker's Guide to the Galaxy, The
    (book series), 241

hockey, 45, 179–180
homicide rates, 107–109
hormone replacement therapy, 122–123
horoscopes, 1, 5, 214–234
horseshoes, 12
Hoxne Hoard, 16–17
Hoyt, Waneta, 210
humanism, secular, 261, 263
Hurricane Joaquin, 260
Hyman, Ray, 247

"I Feel Lucky" (song), 148
improvisational comedy, 165–167
Inhofe, Jim, 125
insurance, 135–136
International Affective Picture System
    (IAPS), 243
Ireland, 194, 258

Jackson, Joe, 36
jade, 13
Jakubczyk-Eckert, Michaela, 129
Japan, 17–19
Jaws (movie), 108
Jesus, 25
Jewel (musician), 198
Johnson, Andrew, 87
Johnson, Lyndon, 87
Jolie, Angelina, 260
Jones, Doug, 190
Jones, Van, 188
Journal of General Psychology, 221
*Journal of Personality and Social
    Psychology*, 242
Journal of Psychology, 221
Judas Iscariot, 12, 25
juggling, 168–169
justice system, 203–213
just luck, 118–119, 277

Kansas City Royals (baseball team), 178

Keaton, Diane, 259

Keeler, Willie, 180

Kelly, Gene, 70

Kennedy, John F., 87–90

King, B.B., 198

Kinsella, W.P., 36

Kirk, Captain, 193

Knights Templar, 25

knock on wood (expression), 12

Kokura, Japan, 17–20, 33

Ku Klux Klan, 259

Lancet, The (journal), 144

Landon, Alf, 187

Last Supper, 13, 25

law of attraction, 72–75

law of large numbers, 133

Lawes, Eric, 16

Lawson, Thomas, 25

Leone, Sergio, 21

leprechauns, 194

liberation theology, 259

Lie groups, 161

lightning, 71–72, 147

Lincoln, Abraham, 87–90

Lincoln (car), 87

linear regression, 175

Literary Digest (magazine), 187–188

Little Boy (bomb), 17

Loki, 25

lotteries, 11, 20, 21, 31–32, 33, 47,
62–65, 111, 146–158, 194, 270

lottery retailer fraud scandal, 152–158,
161–162

Lotto 6/49 (lottery), 31, 151

Louis-Dreyfus, Julia, 26

Louisiana, 261

luck, definition of, 2, 3, 277

Luck Factor, The (book), 195

luck traps, 5, 48–54, 277–278

lucky charms, 2, 3, 173, 194, 266

lucky shot (luck trap), 49, 278

Macbeth (play), 3, 7, 37–38, 56–57, 266

magic, 3, 33–47, 72–75, 200–201,
264–268

Malkovich, John, 259

Mandel, Howie, 164

"Maniac" (song), 183

many people (luck trap), 50, 278

many tries (luck trap), 49, 278

Maple Leafs (hockey team). See Toronto
Maple Leafs

March Madness, 173–177

Margaret, Princess, 25

margin of error, 185–186

marriage, same-sex, 261

Markov chain Monte Carlo, 161

Maui, Hawaii, 135

McCain, John, 184

McLuhan, Marshall, 201

mean reversion, 183

measles, mumps, and rubella (MMR)
vaccination, 137

Men Who Stare at Goats, The (book and
movie), 246

Merchant's House Museum, 39

Merry Millionaire (lottery), 21

Miami Dolphins (football team), 182

Ministry of Defence (UK), 249

moon phases, 79

Moore, Roy, 189–190

morality, 258–262

mucus, 141

multiple sclerosis (MS), 142, 144

Multi-State Lottery Association, 147

murder rates, 107–109
music, 165–167

Nagasaki, 18–20, 33, 112
NASA, 126
National Academy of Sciences, 258
National Football League (NFL), 182
National Geographic (magazine), 106
National Lottery (UK), 31, 158
Nature (journal), 218
NCAA basketball, 173–177
Netflix, 266
neural networks, 76
New England Journal of Medicine, 122
New Jersey, 244
New Mexico, 18
New Mexico (basketball team), 176
New Orleans Saints (football team), 182
Newton's laws of motion, 240
New York State, 21, 69, 210
New York Times, 198
New York Yankees (baseball team), 41,
    177–178, 180
Niffenegger, Audrey, 241
Nightingale, Florence, 116–117
Nobel Prize in Literature, 194
Norse myth, 25
North Carolina, 112
nuclear bombs, 17–19, 112–113
Numb3rs (TV show), 232
numerology, 5, 27, 232–233
Nunavut, 82
nurses, 69–70, 116–117, 211–213,
    226–230

Obama, Barack, 184
O'Connor, Flannery, 70
Oklahoma State (basketball team), 176
one-hit wonder, 182

Oregon (basketball team), 176
Oswald, Lee Harvey, 87–90
out of how many principle (luck trap),
    50, 278

pagan spirits, 12
paraskevidekatriaphobia, 25
Parker, Joe, 14
Pasteur, Louis, 139–140, 216
Peanuts (comic strip), 251
PEAR (Princeton Engineering
    Anomalies Research), 248–249
pedestrian deaths, 169–172
Pell, George, 259
Perkins, Tony, 260–261
p-hacking, 143
pheromones, 79
Philadelphia Warriors (basketball team),
    182
Philip IV (French king), 25
phlegm, 141
Pitt, Brad, 260
placebo effect (luck trap), 51, 68, 278
Plummer, Christopher, 26
Poisson clumping, 171
Poisson distribution, 154, 170
polls. *See* public opinion polls
Popoff, Peter, 236–237
Pouchet, Félix-Archimède, 139
Powerball (lottery), 31, 32, 147–148, 150
precognition, 239–243, 247
Princess Margaret. *See* Margaret, Princess
Princeton Engineering Anomalies
    Research (PEAR), 248–249
pseudorandomness, 127–128
psychic phenomena, 5, 9–10, 60–62,
    236–249
psychological factors (luck trap), 51
Psychological Science (journal), 77

public opinion polls, 45, 184–191
p-value, 119–122, 126, 135, 136, 138, 141, 144, 154, 224–225, 277

quantum mechanics, 241–242, 251
Queen Elizabeth II. *See* Elizabeth II, Queen

rabbits' feet, 3, 12
Radcliffe, Daniel, 259
Randi, James, 218, 224–225, 236–237
random luck, 2, 277
random walks, 161
Rangers (baseball team). *See* Texas Rangers
Red Sox (baseball team). *See* Boston Red Sox
regression, 175
regression to the mean, 183
relativity, theory of, 240–242
religion, 250–263
religious violence, 259–260
replication, 79, 122, 123, 132–145, 224, 230–232, 242–243, 248–249
retrocausality, 241
"Rime of the Ancient Mariner, The" (poem), 13
Roberts, John, 22, 199
Rocky Mountains, 112
romance and luck, 11–12, 42–43, 65–66, 119, 121
Ronson, Jon, 246
Roosevelt, Franklin D., 25, 187
Rose, Pete, 181
Rosenthal Fit, 175
Rosenthal, Helen S., 268–270
Rossini, Gioachino, 25
roulette (game), 133
Royals (baseball team). *See* Kansas City Royals

Royal Society (UK), 258
Royal Statistical Society (UK), 117, 209
Ruby, Jack, 89
Russell, Bertrand, 199
Ruth, Babe, 41

Safeminds, 137
Sagan, Carl, 263
SAIC (Science Applications International Corporation), 246–248
salt, 12
same-sex marriage, 261
Sandy Hook school shooting, 260
Sanitary Commission, 117
Sauldsberry, Woodrow Jr., 182
Savastano, Donald, 21
Savelli, Guy, 246
sayings (aphorisms), 192–202
Science Applications International Corporation (SAIC), 246–248
Secret, The (book), 72–75
secular humanism, 261, 263
Seinfeld (TV show), 26
Sembello, Michael, 183
Semmelweis, Ignaz, 139, 216
September (movie), 253
Serendipity (movie), 42
Serenity Prayer, 13, 48, 115, 264
Sessions, Jeff, 189
sexual orientation, 76
Shakespeare, William, 7, 37–38, 266
Shapiro, Marla, 167–169
Shatner, William, 70
Shenandoah National Park, 71
Shoeless Joe (novel), 3, 36
shotgun effect (luck trap), 49, 278
Shyamalan, M. Night, 35
SIDS (sudden infant death syndrome), 208–210

significance, statistical, 119, 277
Silver, Nate, 184, 190
Sisler, George, 181
Sixth Sense, The (movie), 35
sophomore slump, 182
Soviet Union, 246
sports, 40–42, 45, 131–132, 173–183, 195
Stanley Cup, 45, 179
Stargate Project, 246–248
Starr, Ringo, 198
Star Trek (TV show), 46, 193, 241
Star Wars (movies), 235
statistical significance, 119, 277
statistics, 116–130
sudden infant death syndrome. *See* SIDS
Sullivan, Roy, 71
summer of the gun, 108
summer of the shark, 108
superstitions, 4, 7, 9, 12–13, 24–31, 34, 40–41, 44–46, 56–57, 58–60
Swift, Taylor, 26

Tallady, Christine, 16
Tampa Bay (hockey team), 179
testimonials, 73–75, 128–129
Texas Rangers (baseball team), 178
theft, 105–107, 109–110, 205–207
Thirteen Club, 27
Thirteen Conversations About One Thing (movie), 253
thirteen, fear of, 24–32
Time Machine, The (novel), 241
time travel, 240–241
Time Traveler's Wife, The (novel and movie), 241
Tinkham, Kim, 75
Tipton, Eddie, 147

to multiply or not to multiply (luck trap), 53, 278
Toronto Blue Jays (baseball team), 177–179
Toronto Maple Leafs (hockey team), 45, 179–180
touch wood (expression), 12
traps, luck, 48, 277
Trinity (bomb), 18
true skill, 49
TruMedia Networks, 195
Trump, Donald, 45–46, 107–109, 125, 188–190
tsunamis, 19–20, 34
Twain, Shania, 198
twins, 244–245

United Kingdom (UK), 16–17, 31, 137, 158, 184–185, 208–210, 249
utility functions, 164
Utley, Chase, 181
Utts, Jessica, 247–249

vaccination, 137
Vatican, 259
Virginia, 71

Wakefield, Andrew, 137
Wall Street Journal, 123
warranties, extended, 135, 136
Washington, D.C., 260
Washington, Denzel, 72
Weird or What? (TV show), 70
Wells, H.G., 241
Wettlaufer, Elizabeth, 211
Williams, Robin, 246
Winfrey, Oprah, 72–75
Wiseman, Richard, 195
wishbone, 13, 44

wishes, 44–45

Wizard of Oz, The (movie), 35

Women's Health Initiative, 122–123

wood, knock on (or touch) (expression),
  12

World Series (baseball), 40–42, 177, 180

World War II, 17–20, 22, 35, 195–196

wormholes, 240

Wright, Steven, 246

Yankees (baseball team). *See* New York
  Yankees

Yar, Tasha, 46

Young, Robert, 75

Zamboni, Paolo, 142